Lecture Notes in Mathematics

Edited by A. Dold and B. Eckmann

877

Joachim Erven
Bernd-Jürgen Falkows'.

Low Order Cohomology and Applications

Springer-Verlag
Berlin Heidelberg New York 1981

Authors

Joachim Erven
Siemens AG/Forschungslaboratorien
Otto-Hahn-Ring 6, 8000 München 83
Federal Republic of Germany

Bernd-Jürgen Falkowski
Hochschule der Bundeswehr München, FB Informatik
Werner-Heisenberg-Weg 39, 8014 Neubiberg
Federal Republic of Germany

AMS Subject Classifications (1980): 22-XX

ISBN 3-540-10864-5 Springer-Verlag Berlin Heidelberg New York
ISBN 0-387-10864-5 Springer-Verlag New York Heidelberg Berlin

Printing and binding: Beltz Offsetdruck, Hemsbach/Bergstr.
2141/3140-543210

CONTENTS

INTRODUCTION

Recently continuous tensor products, infinitely divisible positive definite functions and factorizable representations of "current groups" have received much attention. These apparently very different concepts are connected by the fact that the solution of the relevant problems depends on the knowledge of certain cohomology groups. This was first established by Araki [1] for infinitely divisible positive functions and factorizable representations. The probabilistic aspect of the theory and continuous tensor products were described in [20] and also in [6] .

In chapter I we review the relevant definitions of low order cohomology illustrating them by means of some examples of group extensions.

In chapter II we first give a description of continuous tensor products closely following [26] since this seems to be the most intuitive approach (making use of some facts which have become known since 1968). We also exhibit the connections between first and second order cocycles and continuous tensor products, infinitely divisible projective representations and factorizable projective representations. We give realizations of some of these representations in Fock space.

Chapter III is devoted to the computation of cohomology groups for certain semi-direct products (using the Mackey theory of induced representations). Amongst the examples presented are the Euclidean motion groups and the " Leibniz-Extension" of $SL(2;\mathbb{R})$. These results appear to be new.

The whole of chapter IV is needed for the solution of the cocycle problem for $SL(2;\mathbb{R})$. In order to solve the problem it proved necessary to derive all representations of $SL(2;\mathbb{R})$ as induced representations from <u>one</u> subgroup. Again the explicit computation of the cocycles appears to be new. A result on the dimension of the cohomology group is given in [3], but the proof seems to be incomplete. At the end of the chapter we give the corresponding result for $SL(2;\mathbb{C})$ computed in [7].

In chapter V we give some powerful theoretical results. These are essentially contained in [32] and [23] . One first needs to prove Kazdan's result and some results on spherical functions in order to see that only the cohomologies of $SU(n;1)$ and $SO(n;1)$ are really of interest. In order to get precise statements about the dimension of these cohomology groups the connection between Lie Algebra and Lie Group is exploited. (Note that the case of SU(1;1) is explicitly excluded here). We just give some indications as to the method of proof and explain the connection with our concrete calculations in chapter IV.

In chapter VI we deal with genuine infinitely divisible representations (thus solving some technical problems). The explicit formulae for these representations are given using the cohomology groups computed earlier.

By means of two examples we show that even representations constructed from non-trivial cocycles are not always irreducible. Here the example of $SO(3) Ⓢ \mathbb{R}^3$ appears to be new.

In the appendix finally we give the derivation of some results concerning "σ-positive functions" and projective representations which are known but don't seem to be easily accessible in the literature.

It should be mentioned that there appear to be some new results on irreducible representations of current groups (as pointed out by the referee to us). However, at the time of writing, no preprints seem to be available.

Acknowledgement

The authors thank Dr. K. Schmidt (University of Warwick) for several helpful discussions and for suggesting that these notes should be written up at all. The authors also wish to thank Frau B. Leischner and Frl. H. Höhn for the speedy and efficient typing of the final version of this manuscript.

Remark

Most of the results of chapter IV are contained in [4] .

I. CONTINUOUS COHOMOLOGY OF LIE GROUPS AND LIE ALGEBRAS

1. Basic Definitions

The situation which we are going to consider may be described in purely algebraic terms as follows:
We have a sequence of abelian groups $\{C^q\}$ and homomorphisms $\{\delta^q\}$. with the property

$$C^o \xrightarrow{\delta^o} C^1 \xrightarrow{\delta^1} C^2 \xrightarrow{\delta^2} \dots.$$

where $\delta^{q+1} \circ \delta^q = 0$. We now have two distinguished sequences of subgroups of C^q namely

$\{Z^q\}$ where $Z^q := \operatorname{Ker} \delta^q$

$\{B^q\}$ where $B^q := \operatorname{Im} \delta^{q-1}$

Obviously the condition $\delta^{q+1} \circ \delta^q = 0$ is equivalent to $B^q \subseteq Z^q$. Thus we obtain another sequence of groups $\{H^q\}$ by $H^q := Z^q/B^q$. Let us now fix the terminology.

(1.1) Definition:

The elements of

C^q are called q-cochains,

Z^q " " q-cocycles,

B^q " " q-coboundaries.

H^q is called the q-cohomology group.

Although we fix the general definition we are going to be interested mainly in the groups H^1 and H^2.
Let us now apply this process to Lie groups. So suppose that G is a Lie Group and M is a left G-module. Then we consider $C^q(G,M)$, the group of all continuous functions $f:\underbrace{G\times\dots\times G}_{q\text{-times}} \longrightarrow M$ where addition is defined pointwise. This will give the q-dimensional cochain groups.

We further define the coboundary homomorphism

$$\delta^q : C^q(G,M) \longrightarrow C^{q+1}(G,M)$$
$$f \longmapsto \delta f$$

by $$(\delta^q f)(g_1,\ldots,g_{q+1}) := g_1 f(g_2,g_3,\ldots,g_{q+1})$$

$$+ \sum_{i=1}^{q} (-1)^i f(g_1,\ldots,g_i g_{i+1},\ldots,g_{q+1}) + (-1)^{q+1} f(g_1,g_2,\ldots,g_q) .$$

One easily checks by computation that we have $\delta^{q+1} \circ \delta^q = 0$. The resulting cohomology groups are, of course, denoted by $H^q(G,M)$. It should be noted that the action of G on M is part of the definition and a change in the action will in general result in a change of the groups $H^q(G,M)$. It remains to introduce the convention

$$C^0(G,M) := M$$
$$B^0(G,M) := 0 .$$

Since, as mentioned before, we are going to be chiefly interested in the groups $H^1(G,M)$ and $H^2(G,M)$ we are going to look at these in more detail. More precisely still we are going to be interested in $H^1(G,M)$ where M is a seperable Hilbert space $\mathcal{H}$ and the G-action on $\mathcal{H}$ is described by a unitary (orthogonal) representation of G in $\mathcal{H}$. In the case of $H^2(G,M)$ we wish to specialize in the sense that M should be S^1 or $\mathbb{R}$ and the G-action should be trivial. Under these assumptions we then easily obtain the following descriptions for

$B^1(G,\mathcal{H})$: Let $g \mapsto U_g$ be the representation of G in $\mathcal{H}$ under consideration. A 0-cochain is then, by definition, an element $v \in \mathcal{H}$. Thus $(\delta^0 v)(g) = U_g v - v$. Thus the elements of $B^1(G,\mathcal{H})$ are precisely the continuous functions

$f: G \longrightarrow \mathcal{H}$ given by

$f(g) = U_g v - v$ for some $v \in \mathcal{H}$.

$Z^1(G,\mathcal{H})$: A 1-cochain $f \in C^1(G,\mathcal{H})$ is a continuous function $f: G \longrightarrow \mathcal{H}$ while its coboundary is

$$(\delta^1 f)(g_1,g_2) = U_{g_1} f(g_2) - f(g_1 g_2) + f(g_1) .$$

Thus the elements of $Z^1(G,\mathcal{H})$ are precisely the continuous functions $f:G\to\mathcal{H}$ satisfying

$$f(g_1g_2) = U_{g_1}f(g_2)+f(g_1) \ .$$

$\underline{H^1(G,\mathcal{H})}$: This is now $Z^1(G,\mathcal{H})/B^1(G,\mathcal{H})$.
We note here that if the G-action is trivial in $\mathcal{H}$ then $H^1(G,\mathcal{H})$ is just the group of continuous homomorphisms from G to $\mathcal{H}$.

$\underline{B^2(G,\mathbb{R})}$: The elements of $B^2(G,\mathbb{R})$ are just the continuous functions $f:G\times G\to\mathbb{R}$ of the form

$$f(g_1,g_2) = a(g_1)+a(g_2)-a(g_1g_2)$$

where $a:G\to\mathbb{R}$ is just some continuous function.

$\underline{Z^2(G,\mathbb{R})}$: The elements of $Z^2(G,\mathbb{R})$ are the continuous functions $f:G\times G\to\mathbb{R}$ satisfying

$$f(g_1,g_2)+f(g_1g_2,g_3) = f(g_1,g_2g_3)+f(g_2,g_3) \qquad \forall\, g_1,g_2,g_3\in G \ .$$

$\underline{H^2(G,\mathbb{R})}$: This is again $Z^2(G,\mathbb{R})/B^2(G,\mathbb{R})$.

$\underline{B^2(G,S^1)}$: The elements of $B^2(G,S^1)$ are the continuous functions $f:G\times G\to S^1$ of the form

$$f(g_1,g_2) = b(g_1)b(g_2)\overline{b(g_1g_2)}$$

where $b:G\to S^1$ is some continuous function.

$\underline{Z^2(G,S^1)}$: The elements of $Z^2(G,S^1)$ are the continuous functions $f:G\times G\to S^1$ satisfying

$$f(g_1,g_2)f(g_1g_2,g_3) = f(g_1,g_2g_3)f(g_2,g_3) \qquad \forall\, g_1,g_2,g_3\in G \ .$$

$\underline{H^2(G,S^1)}$: This is again $Z^2(G,S^1)/B^2(G,S^1)$.

We now turn to the consideration of Lie algebras. Again the general situation mentioned at the beginning of this section still applies.

However, the cochains and cohomology groups will now be defined more precisely. So let $\mathfrak{g}$ be a Lie algebra and let M be a topological vector space over a field F. The action of the Lie algebra $\mathfrak{g}$ in M will be given by a continuous representation $Q:\mathfrak{g}\to \mathrm{End}(M)$ such that $Q[X,Y] = Q(X)Q(Y) - Q(Y)Q(X)$. Then we consider $C^q(\mathfrak{g},M):=\mathrm{Alt}^q(\mathfrak{g},M)$, the group of continuous alternating multilinear forms of q variables, with addition defined pointwise. These will be the q-cochains in this case.

We further define the coboundary homomorphisms

$$(\delta^q f)(X_1,\ldots,X_{q+1}) :=$$

$$\sum_{i<j} f([X_i,X_j],X_1,\ldots,\hat{X}_i,\ldots,\hat{X}_j,\ldots,X_{q+1})$$

$$+\sum_{i=1}^{q+1} (-1)^i Q(X_i) f(X_1,\ldots,\hat{X}_i,\ldots,X_{q+1})$$

(here "$\wedge$" denotes that the corresponding element should be omitted.)

In view of our applications we are only going to consider the first cohomology group in more detail.

<u>$B^1(\mathfrak{g},M)$:</u> These are all continuous linear functions $f:\mathfrak{g}\to M$ satisfying

$$f(X) = Q(X)v \quad \text{for some fixed } v\in M .$$

<u>$Z^1(\mathfrak{g},M)$:</u> These are all continuous linear functions $f:\mathfrak{g}\to M$ satisfying

$$f([X,Y]) = Q(X)f(Y)-Q(Y)f(X) \quad \forall\, X,Y \in \mathfrak{g}.$$

<u>$H^1(\mathfrak{g},M)$:</u> This again is now just the

quotient $Z^1(\mathfrak{g},M)/B^1(\mathfrak{g},M)$.

2. Some Applications of H^1

In this section we are going to consider some extensions which will clarify the role of H^1 in the cases which we are going to consider. These examples are going to be relevant at a later stage.

Let G be a connected, semi-simple Lie group. Let Π be a continuous representation of G in a Banach space B*). Let the Iwasawa decomposition of G be given by KAN with K compact, A abelian, and N nilpotent. An analytic vector v (with respect to Π) will be called K-finite, if $\Pi(K)v$ generates a finite-dimensional vector space. Let B_K be the space of all analytic K-finite vectors. Finally let $\mathfrak{g}$ be the Lie algebra of G and let $d\Pi$ be the derived representation (of Π). Then B_K is dense in B and invariant under $d\Pi$.

$d\Pi$ is, of course, defined on B_K by

$$d\Pi(X)v: = \frac{d}{dt}\{\Pi(\exp tX)v\}_{t=0} \qquad \forall\, X,v \in \mathfrak{g} \times B_K .$$

We are now able to describe the above-mentioned extensions.

(2.1) Definition:

The generalized Leibnitz extension $G_1(\mathfrak{g}_1)$ of $G(\mathfrak{g})$ is given by

$$G_1: = G \times B \qquad \text{as a set.}$$
$$\mathfrak{g}_1: = \mathfrak{g} \times B_K$$

Group "multiplication" respectively Lie bracket are given by

$$(g_1,v_1)\cdot(g_2,v_2): = (g_1g_2,v_1+\Pi(g_1)v_2)$$

respectively $[(X_1,v_1),(X_2,v_2)]: = ([X_1,X_2],d\Pi(X_1)v_2-d\Pi(X_2)v_1)$.

One verifies easily that G_1 is a group and $\mathfrak{g}_1$ a Lie algebra.
In order to justify the notation as well as to clarify the connection between $\mathfrak{g}_1$ and G_1 we give a sort of "exponential function" from $\mathfrak{g}_1$ to G_1 :

*) We denote group representations by $g \mapsto \Pi(g)$ or $g \mapsto \Pi_g$ which ever is more convenient for readability.

(2.2) Definition:

EXP: $\mathfrak{g}_1 \to G_1$ is defined by

$$\mathrm{EXP}: (X,v) \longmapsto (\exp X, \int_0^1 e^{t d\Pi (X)} v \, dt) .$$

The integral here is to be interpreted as the usual integral for a vector-valued function. The existence is obviously guaranteed. Our terminology is now justified by

(2.3) Lemma:

$t \longmapsto \mathrm{EXP}\ t(X,V)$ is a continuous 1-parameter group with generator (X,v) .

Proof:

$$\mathrm{EXP}\ t(X,v) = (\exp tX, \int_0^t e^{t^1 d\Pi (X)} v \, dt^1) .$$

Obviously we have that $t \longmapsto \mathrm{EXP}\ t(X,V)$ is continuous with

$$\frac{d}{dt}\{\mathrm{EXP}\ t(X,v)\}_{t=0} = (X,v) .$$

Furthermore we obtain

$$\mathrm{EXP}\ t(X,v) \cdot \mathrm{EXP}\ s(X,v) =$$

$$(\exp tX, \int_0^t e^{t^1 d\Pi (X)} v \, dt^1) \cdot (\exp sX, \int_0^s e^{t^1 d\Pi (X)} v \, dt^1) =$$

$$(\exp(s+t)X, \int_0^t e^{t^1 d\Pi (X)} v \, dt^1 + \Pi(\exp tX) \int_0^s \Pi(\exp t^1 X) v \, dt^1)$$

(since v is an analytic vector!)

$$= (\exp(s+t)X, \int_0^t e^{t^1 d\Pi (X)} v \, dt^1 + \int_0^s \Pi (\exp(t+t^1)X) v \, dt^1)$$

$$= (\exp(s+t)X, \int_0^t e^{t^1 d\Pi (X)} v \, dt^1 + \int_0^s e^{(t+t^1) d\Pi (X)} v \, dt)$$

$$= (\exp(s+t)X, \int_0^{s+t} e^{t^1 d\Pi (X)} v \, dt^1)$$

$$= \mathrm{EXP}\ [(t+s)(X,v)] .$$

q.e.d.

We are now going to clarify the role of the cocycles. Suppose we are given embeddings

$$d\phi : \mathfrak{g} \to \mathfrak{g}_1, \quad \phi : G \to G_1 \quad \text{with}$$

$$d\phi : X \mapsto (X, \eta(X)) \quad \text{with } \eta \text{ linear and continuous}$$

$$\phi : g \mapsto (g, \delta(g)) \quad \text{with } \delta \text{ continuous}$$

then we obtain by a fairly trivial computation:

(2.4) Lemma:

(i) $d\,\phi$ is an injective Lie algebra morphism $\Leftrightarrow$ η is a first order cocycle associated with $d\Pi$.

(ii) ϕ is an injective group homomorphism $\Leftrightarrow$ δ is a first order cocycle associated with Π .

So the "number" of possible embeddings of $\mathfrak{g}(G)$ in $\mathfrak{g}_1(G_1)$ is determined by the "size" of the corresponding cohomology group.
We did suggest through the above notation that $d\,\phi$ might be considered as a sort of "differential" for ϕ . The justification for this is provided by:

(2.5) Lemma:

Let η be a cocycle associated with $d\Pi$ and $d\phi : \mathfrak{g} \to \mathfrak{g}_1$ be given by

$$d\,\phi : X \mapsto (X, \eta(X)) \ .$$

Let G be simply connected, let ϕ be chosen such that the following diagram commutes:

$$\begin{array}{ccc} \mathfrak{g} & \xrightarrow{d\phi} & \mathfrak{g}_1 \\ \exp \downarrow & & \downarrow \text{EXP} \\ G & \xrightarrow{\phi} & G_1 \end{array}$$

Then ϕ is an injective group homomorphism.

Proof:

It follows from prop. 5 in [23] that $\delta(\exp X) := \int_0^1 e^{t d\Pi(X)} \eta(x)\,dt$ defines a first order cocycle for Π. The statement of the lemma then follows by applying (2.4). q.e.d.

From the proof of (2.5) it follows immediately that there is a close connection between the Lie algebra cohomology and the cohomology of the Lie group.

Because of its importance later on we wish to consider a special case of $G_1(\mathfrak{g}_1)$. We consider the representation of $\mathfrak{g}$ in $\mathfrak{g}$ defined by $X \mapsto \text{ad}\, X = d\Pi(X)$ with

$$\text{ad}\, X(Y) := [X,Y] \qquad \forall\, X,Y \in \mathfrak{g}$$

Then we obtain as $\mathfrak{g}_1$ the first Leibnitz extension in the sense of [22]. Obviously in this special case we have:

$B_K = B$, $\mathfrak{g}_1$ is a finite-dimensional Lie algebra, G_1 is the associated Lie group and EXP is indeed the usual exponential mapping. One verifies readily that in this case the action of G in $\mathfrak{g}$ is given by

$$\Pi(\exp X)Y = e^{\text{ad}\, X}(Y) .$$

Thus we are lead to

(2.6) Definition:

The Leibnitz extension $G_L(\mathfrak{g}_L)$ of $G(\mathfrak{g})$ is given by

$$\left.\begin{aligned} G_L &:= G \times \mathfrak{g} \\ \mathfrak{g}_L &:= \mathfrak{g} \times \mathfrak{g} \end{aligned}\right\} \text{ as a set .}$$

Group "multiplication" respectively Lie bracket are defined by

$$(g_1,X_1)\cdot(g_2,X_2) := (g_1 g_2, X_1 + \text{Ad}g_1(X_2))$$

respectively $[(X_1,X_1^1),(X_2,X_2^1)] := ([X_1,X_2],[X_1,X_2^1]+[X_1^1,X_2])$

where $\text{Ad}g_1$ is, of course, the differential of the mapping $g \mapsto g_1 g g_1^{-1}$ evaluated at the identity.

It is perhaps interesting to note that in this special case it is possible to give the "number" of embeddings of $\mathfrak{g}$ in $\mathfrak{g}_L$ explicitly:

(2.7) Lemma:

Let $d\phi : \mathfrak{g} \to \mathfrak{g}_L$, defined by $d\phi : X \mapsto (X, \eta(X))$, be a Lie algebra morphism with η linear. Then $\eta(X) = \mathrm{ad}X(X_0)$ for some fixed $X_0 \in \mathfrak{g}$.

Proof:
One verifies readily that η satisfies

$$\eta([X,Y]) = [\eta(X),Y] + [X,\eta(Y)] \qquad X,Y \in \mathfrak{g}.$$

Thus η is a derivation. Since $\mathfrak{g}$ is semi-simple by assumption we know that every derivation must be an inner derivation (cf. [12] p. 122).

q.e.d.

Remarks:

(i) It follows from the statement of (2.7) that the cohomology group associated with the ad-action is trivial!

(ii) Analogous results are valid for the group itself. This is an immediate consequence of the results in [23].

3. An application of H^2

We are here going to describe an extension of a group G which will be used in the appendix. So suppose we are given an element $\sigma \in H^2(G,S^1)$ (with the trivial G-action) satisfying the normalization condition $\sigma(g,e) = \sigma(e,g) = 1 \quad \forall g \in G$. Then we may construct an extension G_σ of G which is described as follows:

$$G_\sigma := G \times S^1 \quad \text{as a set}$$

with group "multiplication" given by

$$(g_1,\lambda_1)\cdot(g_2,\lambda_2) := (g_1 g_2, \lambda_1 \lambda_2 \sigma(g_1,g_2)) \ .$$

The fact that $\sigma \in H^2$ (as described above) ensures that we do indeed obtain a group. In this case again we may try to embed G in G_σ homomorphically by $\phi: g \mapsto (g, a(g))$ say, where

$a: G \to S^1$ is some continuous function.

The condition that ϕ should be a homomorphism then gives:

$$\sigma(g_1, g_2) = \bar{a}(g_1)\bar{a}(g_2)a(g_1g_2) .$$

Thus it is clear that such a homomorphic embedding in this case is possible iff σ is a coboundary.

This extension is fairly important in the theory of projective representations. We shall, however, go into more detail in the appendix.

It should be noted perhaps that we have by no means discussed the topic of continuous cohomology in full generality. We have rather restricted ourselves to cases which are of particular interest to us and only given a sketch of the general background. For more information on this the reader is referred to [31].

In the next chapter we are going to introduce the concept of the continuous tensor product which, at first sight, bears no relation to cohomology at all. We promise, however, that the connection will become obvious fairly soon.

II. CONTINUOUS TENSOR PRODUCTS, INFINITELY DIVISIBLE AND FACTORIZABLE REPRESENTATIONS

1. Continuous Tensor Products (CTPs)

There are several ways to define CTPs (see e.g. [26], [11]). We shall, however, closely follow Streater's method [26] since this construction seems to be the most natural one. Thus let us start with a description of von Neumann's definition of a countable product of Hilbert spaces.

(1.1) Von Neumann's Product

Let $\{\mathcal{H}_i\}_{i\in\mathbb{N}}$ be a sequence of Hilbert spaces and $\Omega=\{\Omega_i\}_{i\in\mathbb{N}}$ be a sequence of unit vectors with $\Omega_i\in\mathcal{H}_i$ for all $i\in\mathbb{N}$. Let $\psi=\{\psi_i\}_{i\in\mathbb{N}}$ be a sequence which differs from Ω in only finitely many places. Suppose further that D is the set of finite formal linear combination of such ψ's. Then we equip D with the following sesquilinear form:

Set $\langle\psi^{(1)},\psi^{(2)}\rangle := \prod_{i=1}^{\infty}\langle\psi_i^{(1)},\psi_i^{(2)}\rangle_{\mathcal{H}_i}$

and extend by linearity/antilinearity. Note that the product always has a finite number of factors!

Obviously $\langle..,..\rangle$ is semi-definite. Separation and completion then gives the required Hilbert space which is, of course, dependent on the "reference vector" Ω .

In order to generalize this construction to a continuous product we need the analogue of the inner product

$$\prod_{i=1}^{\infty}\langle\psi_i^{(1)},\psi_i^{(2)}\rangle_{\mathcal{H}_i} = \exp\left\{\sum_{i=1}^{\infty}\log\langle\psi_i^{(1)},\psi_i^{(2)}\rangle_{\mathcal{H}_i}\right\}$$

where it is fairly obvious that the appearance of the logarithm on the right-hand side of the above equation is going to cause problems. However, the idea is extremely tempting because of its simplicity and indeed, as we shall see, the mentioned problems can, at least in a reasonable number of cases, be resolved.

(1.2) The CTP cf. [26]

We shall try to construct a CTP of copies of the same Hilbert space. Thus let us consider $\{\mathcal{H}_x\}_{x\in\mathbb{R}}$ with $\mathcal{H}_x = \mathcal{H}$ for all $x \in \mathbb{R}$. Suppose that $x \mapsto \Omega_x$ is a continuous map from $\mathbb{R}$ to the unit vectors in $\mathcal{H}$. (We may consider Ω as a section of the Hilbert bundle with total space $\mathbb{R}\times\mathcal{H}$.) Ω is again going to define a reference vector and the analogues to (1.1) should be obvious from the notation. It will be necessary to define for each $x \in \mathbb{R}$ a total set $\Delta_x \subseteq \mathcal{H}_x$ with $\Omega_x \in \Delta_x$ for all $x \in \mathbb{R}$. Then let D be the set of finite formal linear combinations of continuous cross-section ψ which differ from Ω only on a compact set such that $\psi(x) \in \Delta_x \ \forall\, x \in \mathbb{R}$. We intend to furnish D with a sesquilinear semi-definite form and then construct a Hilbert space $\mathcal{H}$ as in (1.1). It should be noted that $\mathcal{H}$ depends on Ω and Δ in general and we shall denote it by $\bigotimes_x^{\Omega,\Delta} \mathcal{H}_x$. We define

$$\langle\psi_1,\psi_2\rangle := \exp\{\int_{\mathbb{R}} \log \langle\psi_1(x),\psi_2(x)\rangle\, dx\} .$$

We say that $\bigotimes_x^{\Omega,\Delta}\mathcal{H}_x$ exists if

(i) $\langle\psi_1(x),\psi_2(x)\rangle \neq 0$ for $\psi_i(x) \in \Delta_x \ \forall\, x \in \mathbb{R}$

(ii) a unique branch of the logarithm can be defined by continuity

(iii) $\sum_{i,j} \alpha_i\bar{\alpha}_j \langle\psi_i,\psi_j\rangle \geq 0 \qquad \alpha_i \in \mathbb{C}$.

Although it is not at all obvious at this stage whether conditions (i)-(iii) can be fulfilled we leave aside these questions for the time being and proceed to give an example.

(1.3) The CTP using projective representations

In [26] Streater gives the example of a CTP of "genuine" representations. Here we use projective representations which seems to lead to some complications of a purely technical nature. However, the reason for our procedure will become obvious once we start discussing questions of existence.

So let G be a Lie Group and suppose that (U,σ,Ω) is a cyclic unitary, projective representation of G in a separable Hilbert space $\mathcal{H}$, i.e.

(i) σ is a continuous second order cocycle ($\sigma \in H^2(G,S^1)$) with $\sigma(g,e) \equiv \sigma(e,g) \equiv 1 \quad \forall\, g \in G$

(ii) $U_{g_1 g_2} = \sigma(g_1,g_2) U_{g_1 g_2} \qquad \forall\, g_1,g_2 \in G$

(iii) U is continuous

(iv) $\{U_g\Omega : g \in G\}$ is total.

In order to define cross-sections we consider the set of all C^∞-maps from $\mathbb{R}$ to G which have compact support (i.e. $\{x:\gamma(x) \neq e\}$ is contained in a compact subset of $\mathbb{R}$). Note that this set is a group under pointwise "multiplication" which we shall denote by $C_e^\infty(\mathbb{R},G)$.

As a total set Δ we shall take

$$\Delta := \{U_{g_1} U_{g_2} \ldots U_{g_n} \Omega : g_i \in G \quad \text{for} \quad 1 \le i \le n,\ n \in \mathbb{N}\}$$

and our cross-sections will then be defined as

$$\psi(x) := U_{\gamma_1(x)} \ldots U_{\gamma_n(x)} \Omega \quad \text{where} \quad \gamma_i \in C_e^\infty(\mathbb{R},G) \quad \text{for} \quad 1 \le i \le n .$$

We consider again the set D of finite formal linear combinations of such ψ and, assuming that the CTP exists, we obtain a sesquilinear form on this set by fixing

$$\langle \psi_1, \psi_2 \rangle := \exp \{\int_{\mathbb{R}} \log \langle \psi_1(x), \psi_2(x) \rangle \, dx\}$$

and extending by linearity/antilinearity.

If $\psi_1(x) = U_{\gamma_1(x)} \ldots U_{\gamma_n(x)} \Omega \qquad \gamma_i \in C_e^\infty(\mathbb{R},G)$

$\psi_2(x) = U_{\beta_1(x)} \ldots U_{\beta_m(x)} \Omega \qquad \beta_i \in C_e^\infty(\mathbb{R},G)$

then we immediately obtain

$$\langle \psi_1, \psi_2 \rangle := \exp \{\int_{\mathbb{R}} \log \langle \psi_1(x), \psi_2(x) \rangle \, dx\}$$

$$\exp\{\int_{\mathbb{R}} \log[(\prod_{i=1}^{n-1} \sigma(\gamma_1(x)\ldots\gamma_i(x),\gamma_{i+1}(x)))(\prod_{j=1}^{m-1} \bar{\sigma}(\beta_1(x)\ldots\beta_j(x),\beta_{j+1}(x)))$$

$$\sigma(\beta_m(x)^{-1}\ldots\beta_1(x)^{-1},\gamma_1(x)\ldots\gamma_n(x))\langle U_{\beta_m(x)^{-1}\ldots\beta_1(x)^{-1}\gamma_1(x)\ldots\gamma_n(x)}\Omega,\Omega\rangle]dx\}.$$

It remains to show that under certain, preferably not very restrictive, conditions these definitions make sense. In order to do so we need to digress and give a description of "infinitely divisible projective representations" in terms of first order cocycles. It will turn out that this particular class of representations is the right one for our problem.

2. Infinitely Divisible Projective Representations and First Order Cocycles

The connection between infinitely divisible projective representations and first order cocycles has been extensively studied in [20] and [6]. We restrict ourselves here to an outline of the proof of the most important theorems and the relevant definitions and refer the reader to the cited literature for further details.

(2.1) Definition:

Let (U,σ,Ω) be a cyclic, unitary, projective representation of a topological group G. We say that (U,σ,Ω) is *infinitely divisible* if for all $n \in \mathbb{N}$ there exist cyclic, unitary projective representations $(U^{(n)},\sigma^{(n)},\Omega^{(n)})$ such that

(i) $\underbrace{U^{(n)} \otimes \ldots \otimes U^{(n)}}_{n\text{-times}}$ and U are unitarily equivalent

and Ω and $\underbrace{\Omega^{(n)} \otimes \ldots \otimes \Omega^{(n)}}_{n\text{-times}}$ correspond under this equivalence.

(ii) ${\sigma^{(n)}}^n \equiv \sigma$ (The n-th power is here just the pointwise product!)

The two apparently unrelated concepts of first order cocycle and infinitely divisible projective representation are fixed up by the following

(2.2) Theorem:

Let G be a connected, locally connected, locally compact, second countable group. Let (U,σ,Ω) be an infinitely divisible, cyclic, unitary, projective representation. Then there exists a first order cocycle δ associated with a unitary representation $g \mapsto V_g$ of G such that

(i) $\langle U_g\Omega,\Omega\rangle = \exp\left[-\frac{1}{2}\langle\delta(g),\delta(g)\rangle + ia(g)\right]$

(ii) $\sigma(g_1,g_2) = \exp i[\mathrm{Im}\langle\delta(g_2),\delta(g_1^{-1})\rangle + a(g_1)+a(g_2)-a(g_1g_2)]$

where $a:G \to \mathbb{R}$ is continuous with

$a(g^{-1}) = -a(g) \quad \forall g \in G$.

Proof:
Theorem (3.6) in [6] allows us to transform the given projective representation to a "canonical form". Theorem (3.7) in [6] together with corollary (12.8) in [20] guarantee the existence of the required logarithms. The precise nature of these logarithms is then determined by Theorem (3.1) in [6]. q.e.d.

The above characterization of the *"expectation value"* $\langle U_g\Omega,\Omega\rangle$ of an infinitely divisible projective representation and its associated *"multiplier"* σ in terms of first order cocycles is going to prove very useful as we shall see in the next section.

3. Necessary and Sufficient Conditions for the Existence of a CTP of Projective Representations

In this section we are going to provide an answer for some of the questions arising at the end of section 1.

(3.1) Theorem:

A necessary and sufficient condition for the existence of the CTP described in (1.3) is that the given projective representation (U,σ,Ω) is infinitely divisible. (Note that we assume here that the conditions of (2.2) for G are satisfied!)

Proof:
First assume that (U,σ,Ω) is infinitely divisible. An application of (2.2) yields

$$\langle U_g\Omega,\Omega\rangle = \exp\left[-\frac{1}{2}\langle\delta(g),\delta(g)\rangle + ia(g)\right] ,$$

where a,δ are as in (2.2). Thus

if $\psi_1(x) = U_{\gamma_1(x)}\cdots U_{\gamma_n(x)}\Omega \qquad \gamma_i \in C_e^\infty(\mathbb{R},G)$

$\psi_2(x) = U_{\beta_1(x)}\cdots U_{\beta_n(x)}\Omega \qquad \beta_j \in C_e^\infty(\mathbb{R},G)$

then we obtain for our sesquilinear form:

$$\langle\psi_1,\psi_2\rangle: =$$

$$\exp\int_{\mathbb{R}}\Big[\sum_{i=1}^{n-1} is(\gamma_1(x)\ldots\gamma_i(x),\gamma_{i+1}(x)) - \sum_{j=1}^{m-1} is(\beta_1(x)\ldots\beta_j(x),\beta_{j+1}(x))$$

$$+ is(\beta_m(x)^{-1}\ldots\beta_1(x)^{-1},\gamma_1(x)\ldots\gamma_n(x))$$

$$- \frac{1}{2}\langle\delta(\beta_m(x)^{-1}\ldots\beta_1(x)^{-1}\gamma_1(x)\ldots\gamma_n(x)),\delta(\beta_m(x)^{-1}\ldots\gamma_n(x))\rangle$$

$$+ ia(\beta_m(x)^{-1}\ldots\beta_1(x)^{-1}\gamma_1(x)\ldots\gamma_n(x))\Big]dx$$

where we have set

$$s(g_1,g_2): = \operatorname{Im}\langle\delta(g_2),\delta(g_1^{-1})\rangle + a(g_1) + a(g_2) - a(g_1g_2) .$$

We need to show that our sesquilinear form is positive semi-definite. To this end we consider direct integrals. Suppose that δ is associated with a representation $g \mapsto V_g$.

Let $\Delta(\gamma) := \int^{\oplus} \delta^x(\gamma(x))dx$ with $\delta^x \equiv \delta \quad \forall x \in \mathbb{R}$

$V_0(\gamma) := \int^{\oplus} V^x_{\gamma(x)} dx$ with $V^x \equiv V \quad \forall x \in \mathbb{R}$.

Then it is clear that we have

$$V_0(\gamma_1)\Delta(\gamma_2) = \Delta(\gamma_1\gamma_2) - \Delta(\gamma_1) .$$

A fairly tedious (but straightforward) calculation now gives:

$$\langle\psi_1,\psi_2\rangle = \exp \alpha(\gamma_1\ldots\gamma_n) \exp \bar{\alpha}(\beta_1\ldots\beta_m) \exp \langle\Delta(\gamma_1\ldots\gamma_n),\Delta(\beta_1\ldots\beta_m)\rangle$$

where

$$\alpha(\gamma_1\ldots\gamma_n) :=$$

$$i\int_{\mathbb{R}}[\sum_{i=1}^{n-1} s(\gamma_1(x)\ldots\gamma_i(x),\gamma_{i+1}(x)) + a(\gamma_1(x)\ldots\gamma_n(x))]dx - \frac{1}{2}\|\Delta(\gamma_1\ldots\gamma_n)\|^2 .$$

(Note that all "products" of functions are to be taken pointwise!)

Using the fact that the exponential of a positive semi-definite "kernel" is again positive semi-definite one easily obtains that the given sesquilinear form is semi-definite. Separation and completion then give the required CTP. Thus we have shown the sufficiency of our condition. In order to see the necessity of the condition we can adapt the argument given in [26] (Theorem 1) very easily. It must be noted, however, that there is a one-one correspondence between infinitely divisible projective representations (U,σ,Ω) and so-called infinitely, divisible σ-positive functions. Details of this fact may be found in [10].

q.e.d.

Remark: We could have proved sufficiency in (3.1) analoguously to [26] as well. Our proof, however, incorporates a complete characterization of infinitely divisible projective representations in terms of first order cocycles and thus will be of use later on.

4. CTPs of Representations of $C_e^\infty(\mathbb{R},G)$

Up to now we have dealt only with CTPs of Hilbert spaces. With the above construction, however, we are able to obtain a CTP of projective representations of the "current group" $C_e^\infty(\mathbb{R},G)$. Let (U,σ,Ω) be as in section 3. Then for each $x \in \mathbb{R}$ we can define a projective representation U^x of $C_e^\infty(\mathbb{R},G)$ in $\mathcal{H}$ by means of:

$$U_\gamma^x U_{g_1} \cdots U_{g_n} \Omega := U_{\gamma(x)} U_{g_1} \cdots U_{g_n} \Omega \ .$$

Obviously we have:

$$U_{\gamma_1}^x U_{\gamma_2}^x = \sigma(\gamma_1(x),\gamma_2(x)) U_{\gamma_1\gamma_2}^x \ .$$

The continuous tensor product of these representations is now obtained by setting:

$$[\tilde{U}(\gamma)\psi](x) := U_{\gamma(x)} U_{\gamma_1(x)} \cdots U_{\gamma_n(x)} \Omega$$

where $\psi(x) = U_{\gamma_1(x)} \cdots U_{\gamma_n(x)} \Omega$ and $\gamma, \gamma_i \in C_e^\infty(\mathbb{R},G)$.

Referring to section 3 we see that this defines indeed a cyclic projective representation $(\tilde{U},\tilde{\sigma},\tilde{e})$ where

(i) $\tilde{U}(\gamma_1)\tilde{U}(\gamma_2) = \tilde{\sigma}(\gamma_1,\gamma_2)\tilde{U}(\gamma_1\gamma_2)$

(ii) $\tilde{\sigma}(\gamma_1,\gamma_2) = \exp \int_{\mathbb{R}} \log \sigma(\gamma_1(x),\gamma_2(x))dx$

(iii) $\langle\tilde{U}(\gamma)\tilde{e},\tilde{e}\rangle = \exp \int_{\mathbb{R}} \log\langle U_\gamma^x \Omega,\Omega\rangle dx$

$\tilde{e}(x) \equiv \Omega \qquad \forall\, x \in \mathbb{R}$.

Indeed for reference purposes we note that (using (3.1)) (ii) and (iii) are explicitly given by

$$\tilde{\sigma}(\gamma_1,\gamma_2) = \exp i\{\int_{\mathbb{R}} [a(\gamma_1(x)) + a(\gamma_2(x)) - a(\gamma_1\gamma_2(x) + \\ + \operatorname{Im}\langle\delta(\gamma_2(x)),\delta(\gamma_1(x)^{-1})\rangle]dx\}$$

$$\langle\tilde{U}(\gamma)\tilde{e},\tilde{e}\rangle = \exp\{i\int_{\mathbb{R}} [a(\gamma(x)) - \frac{1}{2}\langle\delta(\gamma(x)),\delta(\gamma(x))\rangle]dx\} \ .$$

It should be pointed out that the multiplier in (ii) above has exactly the form which we would expect: Since the multiplier for the tensor product of two projective representations is just the product of the two multipliers we see that (ii) represents the obvious generalization of this concept. We now wish to exhibit the connection between the CTP constructed above and the so-called "factorizable representations" of Current Groups. So we shall digress a little in the following sections.

5. Factorizable Projective Representations of Current Groups and Fock Space

These representations are of interest in Mathematical Physics, see e.g. [17]. We need some definitions from [22].

(5.1) Definition:

$C_e^\infty(\mathbb{R},G)$ with the pointwise "multiplication" and furnished with a Schwartz topology is called the *Current Group*. It is obviously a group.

Remark: Slightly different definitions of Current Groups are also to be found in the literature, cf. [21].

As before we shall only consider continuous, unitary (projective) representations of Current Groups.

For every compact set $K \subseteq \mathbb{R}$ let $C_e^\infty(K,G) \subseteq C_e^\infty(\mathbb{R},G)$ be the subgroup of those C^∞-maps whose support is contained in K. If $K_1, K_2 \subseteq \mathbb{R}$ are disjoint compact subsets of $\mathbb{R}$ then we can identify $C_e^\infty(K_1 \cup K_2, G)$ in a natural manner with $C_e^\infty(K_1,G) \times C_e^\infty(K_2,G)$:
For each $\gamma \in C_e^\infty(K_1 \cup K_2, G)$ we define

$$\gamma_i(x) := \begin{cases} \gamma(x) & x \in K_i \\ e & x \notin K_i \end{cases} \qquad i=1,2 \ .$$

Then we have $\gamma = \gamma_1\gamma_2$ and the map $\gamma \mapsto (\gamma_1, \gamma_2)$ gives the required isomorphism.

Given a representation U of $C_e^\infty(\mathbb{R},G)$ we define a representation U^K for the subgroup $C_e^\infty(K,G)$ by $U_\gamma^K := U_\gamma$ for $\gamma \in C_e^\infty(K,G)$.
This leads to

(5.2) Definition:

A representation U of $C_e^\infty(\mathbb{R},G)$ is called *factorizable* if for any two $K_1, K_2 \subseteq \mathbb{R}$ with $K_1 \cap K_2 = \phi$ and K_1, K_2 compact we have:

$$U^{K_1 \cup K_2} \text{ is unitarily equivalent to } U^{K_1} \otimes U^{K_2} .$$

This unitary equivalence depends, of course, on K_1 and K_2.

We are now going to describe the standard method for constructing factorizable representations and for this reason we are going to sketch some important facts concerning the construction of "Fock Space". For details cf. [11].
Let $\mathcal{H}$ be a (real or complex) Hilbert space and let $\underset{n}{\otimes}\mathcal{H}$ be the n-fold tensor product. Then the symmetric group σ_n acts on $\underset{n}{\otimes}\mathcal{H}$ by means of unitary operators. This action is given by

$$\forall\ S \in \sigma_n \quad U_{S,n}(x_1 \otimes \ldots \otimes x_n) := x_{S(1)} \otimes \ldots \otimes x_{S(n)} .$$

The operator $P_n := \frac{1}{n!} \sum_{S \in \sigma_n} U_{S,n}$ is then an orthogonal projection. Its image (the set of all elements in $\underset{n}{\otimes}\mathcal{H}$, which are invariant under $U_{S,n}$) will be denoted by $S^n\mathcal{H}$ and we shall fix $S^0 := \mathbb{R}$ or $\mathbb{C}$.

(5.3) Definition:

The *Fock Space* $S\mathcal{H}$ over $\mathcal{H}$ is given by $S\mathcal{H} := \bigoplus_{n=\sigma}^{\infty} S^n\mathcal{H}$.

The elements of $S\mathcal{H}$ are sequences (x_n) with $x_n \in S^n\mathcal{H}$ and $\sum_{n=0}^{\infty} ||x_n||^2 < \infty$. The scalar product is defined by $<x,y> := \sum_{n=0}^{\infty} <x_n, y_n>$.
If $\mathcal{H}$ is separable and $\{e_i\}$ an orthonormal basis of $\mathcal{H}$, then we obtain an orthonormal basis for $S^n\mathcal{H}$ as follows:

Let (n_i) be a sequence of natural numbers with $\Sigma n_i = n$, then the elements

$$\left\{\frac{n!}{n_1!n_2!\ldots}\right\}^{1/2} P_n\left[\left(\bigotimes_{n_1} e_1\right) \otimes \left(\bigotimes_{n_2} e_2\right) \otimes \ldots\right]$$

form an orthonormal basis for $S^n\mathcal{H}$.

(5.4) Definition:

For each $a \in \mathcal{H}$ Exp a is defined by:

$$\text{Exp } a := \left(1, a, \frac{1}{\sqrt{2!}}\, a \otimes a, \frac{1}{\sqrt{3!}}\, a \otimes a \otimes a, \ldots\right).$$

(5.5) Lemma:

The following statements are valid:

(i) $\langle \text{Exp } a, \text{Exp } b\rangle = e^{\langle a,b\rangle}$

(ii) $\|\text{Exp } a\| = e^{\frac{\|a\|^2}{2}} \geq \|a\|$

(iii) $|\langle \text{Exp } a, \text{Exp } b\rangle| = e^{\text{Re}\langle a,b\rangle}$.

Proof:
Computation.

(5.6) Lemma:

The mapping $\text{Exp}: \mathcal{H} \to S\mathcal{H}$ is injective and continuous.

Proof:
See [11].

(5.7) Lemma:

The elements Exp a with $a \in \mathcal{H}$ are linearly independent and span $S\mathcal{H}$.

Proof:
See [11].

We return now to the construction of factorizable representations. Since the role of first order cocycles (yes, these will be of importance here, too) comes out more clearly in the case of projective representations we are going to tackle the problem from there. We need a further definition.

Let a projective representation (U,σ) of $C_e^\infty(\mathbb{R},G)$ be given. We then define a projective representation (U^K,σ^K) of $C_e^\infty(K,G)$ (K compact) by

$$U_\gamma^K := U_\gamma \quad \text{for} \quad \gamma \in C_e^\infty(K,G)$$

$$\sigma^K(\gamma_1,\gamma_2) := \sigma(\gamma_1,\gamma_2) \quad \text{for} \quad (\gamma_1,\gamma_2) \in C_e^\infty(K,G) \times C_e^\infty(K,G) .$$

This leads to

(5.8) Definition:

A projective representation (U,σ) of $C_e^\infty(\mathbb{R},G)$ is called *factorizable*, if for any two compact disjoint subsets K_1,K_2 of $\mathbb{R}$ we have

$$(U^{K_1 \cup K_2},\sigma^{K_1 \cup K_2}) \quad \text{is unitarily equivalent to} \quad ((U^{K_1} \otimes U^{K_2}),\sigma^{K_1}\sigma^{K_2}) .$$

Remarks:

1. If $\sigma \equiv 1$ this reduces to (5.2).
2. Slightly more general factorizable representations were analyzed in [20], [21].

Now let $\mathcal{H}$ be a separable, complex Hilbert space. Let Π be a unitary representation of G in $\mathcal{H}$ and let δ be a first order cocycle associated with Π.

We consider the direct integral $\mathcal{H}_0 = \int^\oplus \mathcal{H}_x dm(x)$ with $\mathcal{H}_x = \mathcal{H}$ $\forall x \in \mathbb{R}$ and m some $(\sigma\text{-})$ finite measure on $\mathbb{R}$. We extend δ to a function Δ: $C_e^\infty(R,G) \to \mathcal{H}_0$ and Π to a representation Π_0 of $C_e^\infty(\mathbb{R},G)$ in $\mathcal{H}_0$ by means of

$$\Delta(\gamma) := \int^{\oplus} \delta^{x}(\gamma(x))\,dm(x)$$

$$\Pi_0(\gamma) := \int^{\oplus} \Pi^{x}(\gamma(x))\,dm(x)$$

where we have set $\delta^x(\cdot):=\delta(\cdot), \Pi^x:=\Pi \quad \forall\, x \in \mathbb{R}$. Then Δ as well as Π_0 are direct integrals. We note that for all $\gamma_1, \gamma_2 \in C_e^\infty(\mathbb{R},G)$ we have:

$$\Pi_0(\gamma_1)\Delta(\gamma_2) = \Delta(\gamma_1\gamma_2) - \Delta(\gamma_1) \ .$$

Thus Δ is a first order cocycle associated with Π_0 .

For all $(\gamma_1,\gamma_2) \in C_e^\infty(\mathbb{R},G) \times C_e^\infty(\mathbb{R},G)$ we now define:

$$\varphi(\gamma_1) := -\frac{1}{2}\,\|\Delta(\gamma_1)\|^2 = -\frac{1}{2}\int_{\mathbb{R}} <\delta(\gamma_1(x))\ ,\delta(\gamma_1(x))>dm(x)$$

$$S(\gamma_1,\gamma_2) := \mathrm{Im}\, <\Delta(\gamma_2),\Delta(\gamma_1{}^{-1})>$$

$$= \mathrm{Im}\int_{\mathbb{R}} <\delta(\gamma_2(x)),\delta(\gamma_1(x)^{-1})>dm(x) \ .$$

For each $\gamma \in C_e^\infty(\mathbb{R},G)$ we can now define an operator $U(\gamma)$ on the subspace of $S\mathfrak{X}_0$ which is spanned by $\{\mathrm{Exp}\,\Delta(\gamma) : \gamma \in C_e^\infty(\mathbb{R},G)\}$ by setting

$$U(\gamma) : \mathrm{Exp}\ \Delta(\gamma') \mapsto$$

$$\exp\{is(\gamma,\gamma') + \varphi(\gamma) - \mathrm{Re}<\Pi_0(\gamma),\Delta(\gamma)>\}\ \mathrm{Exp}\ \Delta(\gamma\gamma')$$

and extending to the whole subspace by linearity.
One verifies readily:

(i) $U(\tilde{e}) = I$ (where $\tilde{e}(x) \equiv e \quad \forall\, x \in \mathbb{R}$)

(ii) $U(\gamma_1)U(\gamma_2) = \exp is(\gamma_1,\gamma_2)U(\gamma_1\gamma_2)$

(iii) $U(\gamma)$ is unitary $\forall\, \gamma \in C_e^\infty(\mathbb{R},G)$

(iv) $(U, \exp is)$ is continuous.

Then $(U,\exp is)$ is a projective representation of $C_e^\infty(\mathbb{R},G)$. We note that $\mathrm{Exp}\ \Delta(\tilde{e})$ is a cyclic vector for this representation.

We are now going to show that the projective representation constructed above is indeed factorizable. First of all we have to characterize projective representations with cyclic vectors by their *"expectation values"*.

(5.9) Definition:

The *expectation value* of the representation U (as above) is given by

$$E(U(\gamma)) := \langle U(\gamma)\ \mathrm{Exp}\ \Delta(\tilde{e}),\ \mathrm{Exp}\ \Delta(\tilde{e})\rangle .$$

We recall that as mentioned before projective representations (U,σ) with cyclic vectors are defined up to unitary equivalence by giving σ and their expectation values. The proof can be given by reducing the projective representation to a genuine representation (using a central extension) and then adapting the proof of the GNS construction. For details see the appendix.

(5.10) Lemma:

The representation $(U,\ \exp is,\ \mathrm{Exp}\ \Delta(\tilde{e}))$ described above is factorizable.

Proof:

Let $K_1, K_2 \subseteq \mathbb{R}$ with K_1, K_2 compact and $K_1 \cap K_2 = \emptyset$. Then we have: $(U^{K_1 \cup K_2}, \sigma^{K_1 \cup K_2})$ is a projective representation in the Hilbert space $\mathcal{H}(K_1 \cup K_2)$, which is spanned by $\{\mathrm{Exp}\ \Delta(\gamma) : \gamma \in C_e^\infty(K_1 \cup K_2, G)\}$ with cyclic vector $\mathrm{Exp}\ \Delta(\tilde{e})$. Similarly (U^i, σ^{K_i}), $i=1,2$ are projective representations in $\mathcal{H}(K_i)$ with cyclic vectors $\mathrm{Exp}\ \Delta(\tilde{e})$, whilst $(U^{K_1} \otimes U^{K_2}, \sigma^{K_1}\sigma^{K_2})$ is of course a projective representation in $\mathcal{H}(K_1) \otimes \mathcal{H}(K_2)$ with cyclic vector $\mathrm{Exp}\ \Delta(\tilde{e}) \otimes \mathrm{Exp}\ \Delta(\tilde{e})$. (For brevity we have set $\sigma \equiv \exp is$!)

Suppose now that $\gamma \in C_e^\infty(K_1 \cup K_2, G)$ with $\gamma = \gamma_1\gamma_2$ and $\gamma_i \in C_e^\infty(K_i, G)$. Then we obtain:

$$(U^{K_1 \cup K_2}(\gamma)) = \langle U^{K_1 \cup K_2}(\gamma)\ \mathrm{Exp}\ \Delta(\tilde{e}),\ \mathrm{Exp}\ \Delta(\tilde{e})\rangle = \exp \varphi(\gamma)$$

$$= \exp [- \tfrac{1}{2}\|\Delta(\gamma_1\gamma_2)\|^2]$$

$$= \exp [- \tfrac{1}{2}\|\Pi_O(\gamma_1)\Delta(\gamma_2)+\Delta(\gamma_1)\|^2]$$

$$= \exp [- \tfrac{1}{2} (\|\Delta(\gamma_1)\|^2+\|\Delta(\gamma_2)\|^2]$$

$$= E(U^{K_1}(\gamma_1))\,E(U^{K_2}(\gamma_2))$$

$$= E(U^{K_1}(\gamma_1) \otimes U^{K_2}(\gamma_2)) \ .$$

Suppose further that $\gamma' \in C_e^\infty(K_1 \cup K_2, G)$ with $\gamma' = \gamma_1'\gamma_2'$ and $\gamma_i' \in C_e^\infty(K_i, G)$, $i=1,2$.

Then it follows that

$$\sigma^{K_1 \cup K_2}(\gamma,\gamma') = \exp [iS^{K_1 \cup K_2}(\gamma,\gamma')]$$

$$= \exp \ i \ \mathrm{Im}\langle\Delta(\gamma'),\Delta(\gamma^{-1})\rangle$$

$$= \exp \ i \ \mathrm{Im}[\langle\Delta(\gamma_1'),\Delta(\gamma_1^{-1})\rangle+\langle\Delta(\gamma_2'),\Delta(\gamma_2^{-1})\rangle] \quad \text{as above}$$

$$= \sigma^{K_1}(\gamma_1,\gamma_1')\,\sigma^{K_2}(\gamma_2,\gamma_2') \ .$$

The computations together with the remark made after (5.9) show that

$(U^{K_1 \cup K_2}, \sigma^{K_1 \cup K_2})$ is unitarily equivalent to $(U^{K_1} \otimes U^{K_2}, \sigma^{K_1}\sigma^{K_2})$.

q.e.d.

Remark:

If we construct the projective representation by using a real cocycle (so that $\exp iS \equiv 1$) then we obtain, of course, a genuine factorizable representation. This special case is dealt with in [29] where only real Hilbert spaces are considered. One can, however, obtain genuine factorizable representations without $\exp iS \equiv 1$. Since this leads to a special case we shall discuss this further in the next section.

6. Coboundaries and their Associated Representations

We keep the same notation as in section 5. Let an irreducible representation of G in $\mathcal{H}$ be given (denoted by Π). A coboundary associated with Π is then a special cocycle of the form $\delta(g) = \Pi(g)v - v$ for fixed $v \in \mathcal{H}$ (cf. I.).
In this case we obtain

$$\varphi(\gamma) = -\frac{1}{2}\|\Delta(\gamma)\|^2 = \int_{\mathbb{R}} \mathrm{Re}\, \langle \Pi(\gamma(x))v - v, v \rangle dm(x) .$$

Suppose now that m is a <u>finite</u> measure. U is then defined by

$$U(\gamma): \mathrm{Exp}\, \Delta(\gamma') \mapsto \exp\, [iS(\gamma,\gamma') + \varphi(\gamma\gamma') - \varphi(\gamma')]\, \mathrm{Exp}\, \Delta(\gamma\gamma') .$$

Here S is given by

$$S(\gamma,\gamma') = a(\gamma\gamma') - a(\gamma) - a(\gamma') \quad \text{with}$$

$$a(\gamma) = \int_{\mathbb{R}} \mathrm{Im} \langle \Pi(\gamma(x))v, v \rangle dm(x) .$$

Obviously we still have:

$$U(\gamma)U(\gamma') = \exp\, iS(\gamma,\gamma')U(\gamma\gamma') .$$

Because of the special form of S (S is technically speaking a second order coboundary) we can now construct an associated genuine representation $\gamma \mapsto V(\gamma)$ by setting

$$V(\gamma) := e^{ia(\gamma)}U(\gamma) .$$

We then immediately obtain

(6.1) Lemma:

$\gamma \mapsto V(\gamma)$ is a genuine factorizable representation of $C_e^\infty(\mathbb{R},G)$.

Proof:

(i) $V(\gamma_1)V(\gamma_2) = e^{ia(\gamma_1)}e^{ia(\gamma_2)}U(\gamma_1)U(\gamma_2)$

$$= e^{i[a(\gamma_1)+a(\gamma_2)+a(\gamma_1\gamma_2)-a(\gamma_1)-a(\gamma_2)]} U(\gamma_1\gamma_2) = V(\gamma_1\gamma_2) .$$

(ii) The expectation value $E(V(\gamma))$ (V is obviously cyclic with cyclic vector $\operatorname{Exp} \Delta(\tilde{e})$) is given by

$$E(V(\gamma)) = e^{[ia(\gamma)+\varphi(\gamma)]} .$$

Suppose now that $K_1, K_2 \subseteq \mathbb{R}$ are compact with $K_1 \cap K_2 = \emptyset$ and $\gamma_1 \in C_e^\infty(K_i, G)$ $i=1,2$ then it follows that

$$\begin{aligned} E(V(\gamma_1\gamma_2)) &= e^{[ia(\gamma_1\gamma_2)+\varphi(\gamma_1\gamma_2)]} \\ &= e^{[ia(\gamma_1)+\varphi(\gamma_1)]} e^{[ia(\gamma_2)+\varphi(\gamma_2)]} \\ &= E(V(\gamma_1))E(V(\gamma_2)) . \end{aligned}$$

Thus V is factorizable. q.e.d.

We are now going to describe V , up to unitary equivalence, in terms of Π thus clarifying the situation somewhat.

We consider the direct integral

$$\Pi_0(\gamma) = \int^{\oplus} \Pi^x(\gamma(x))dm(x) \quad \text{acting in}$$

$$\mathcal{H}_0 = \int^{\oplus} \mathcal{H}_x dm(x) \quad (\Pi_x = \Pi, \mathcal{H}_x = \mathcal{H} \quad \forall x \in \mathbb{R}) .$$

Let now $v_{\mathcal{H}_0} := \int^{\oplus} v_x dm(x)$ with $v_x = v$ $\forall x \in \mathbb{R}$. Then we can immediately describe a cyclic representation V' of $C_e^\infty(\mathbb{R},G)$ in the subspace of $S\mathcal{H}_0$ which is spanned by $\{\operatorname{Exp}[\Pi_0(\gamma)v_{\mathcal{H}_0}] : \gamma \in C_e^\infty(\mathbb{R},G)\}$. We set:

$$V'(\gamma) \operatorname{Exp}[\Pi_0(\gamma')v_{\mathcal{H}_0}] := \operatorname{Exp}[\Pi_0(\gamma\gamma')v_{\mathcal{H}_0}] \quad \forall \gamma,\gamma \in C_e^\infty(\mathbb{R},G)$$

and extend by linearity.

Obviously we obtain a cyclic representation with cyclic vector $\operatorname{Exp} v_{\mathcal{H}_0}$. For convenience we are going to normalize the cyclic vector and thus we shall use

$$\exp\left[-\tfrac{1}{2}\|v_{\mathcal{H}_0}\|^2\right] \operatorname{Exp} v_{\mathcal{H}_0}$$

as a cyclic vector.

The connection between our given representation V and the new representation V' is obtained by calculating the expectation value. We have:

$$\begin{aligned} E(V'(\gamma)) &= \exp\left[-\|v_{\mathcal{H}_0}\|^2\right] \langle V'(\gamma) \operatorname{Exp} v_{\mathcal{H}_0}, \operatorname{Exp} v_{\mathcal{H}_0} \rangle \\ &= \exp\left[-\|v_{\mathcal{H}_0}\|^2\right] \langle \operatorname{Exp}\left[\Pi_0(\gamma) v_{\mathcal{H}_0}\right], \operatorname{Exp} v_{\mathcal{H}_0} \rangle \\ &= \exp\left[\int_{\mathbb{R}} \langle \Pi(\gamma(x)) v - v, v \rangle \, dm(x)\right] . \end{aligned}$$

Thus, V and V' have the same expectation value. This leads to

(6.2) Lemma:

V and V' are unitarily equivalent. So we have shown that V' is factorizable also.

Remark:

It is clear from the above that one can associate to any given representation Π a factorizable representation of the current group $C_e^\infty(\mathbb{R},G)$. This representation, however, is <u>reducible</u> since it leaves the so-called n-particle space $S^n\mathcal{H}_0$ invariant.

Above we have described explicitly a construction where the measure appearing in the direct integral is finite. If a σ-finite measure is used this is not quite so easy. (For simplicity we are only going to consider Lebesgue measure.)

First of all we have, as above

$$\delta(g) = \Pi(g)v - v \qquad v \in \mathcal{H} \text{ fixed}$$

$$\varphi(\gamma) = -\frac{1}{2}\|\Delta(\gamma)\|^2 = \int_{\mathbb{R}} \langle \Pi(\gamma(x))v - v, v \rangle dx$$

$$U(\gamma):\mathrm{Exp}\,\Delta(\gamma') \mapsto \exp\,[iS(\gamma,\gamma')+\varphi(\gamma\gamma')-\varphi(\gamma)]\mathrm{Exp}\,\Delta(\gamma\gamma')$$

$$S(\gamma,\gamma') = a(\gamma\gamma')-a(\gamma)-a(\gamma') \quad \text{with}$$

$$a(\gamma) = \int_{\mathbb{R}} \mathrm{Im}\, \langle\Pi(\gamma(x))v,v\rangle dx$$

$$U(\gamma)U(\gamma') = \exp\, iS(\gamma,\gamma')U(\gamma\gamma') \;.$$

Again we construct a cyclic representation from this by setting

$$V(\gamma) := e^{ia(\gamma)}U(\gamma) \;.$$

The expectation value (with respect to the obvious cyclic vector) is then given by

$$E(V(\gamma)) = \exp\,\{\int_{\mathbb{R}}\langle\Pi(\gamma(x))v-v,v\rangle dx\} \;.$$

In order to construct an equivalent representation we consider an enumeration I_i, $i\in\mathbb{N}$, of the half-open unit intervals with $\bigcup_{i=1}^{\infty} I_i=\mathbb{R}$. We define

$$\Pi_i(\gamma) := \int_{I_i}^{\oplus} \Pi^x(\gamma(x))dx \qquad \text{with } \Pi^x \equiv \Pi \quad \forall\, x\in I_i$$

$$_i \;:= \int_{I_i}^{\oplus} \mathcal{H}_x dx \qquad \text{with } \mathcal{H}_x \equiv \mathcal{H} \quad \forall\, x\in I_i$$

$$v_{\mathcal{H}_i} := \int_{I_i}^{\oplus} v_x dx \qquad \text{with } v_x = v \quad \forall\, x\in I_i \;.$$

Suppose that cyclic representations V_i of $C_e^{\infty}(\mathbb{R},G)$ in subspaces L_i of $S\mathcal{H}_i$ spanned by

$$\{\mathrm{Exp}\,[\Pi_i(\gamma)v_{\mathcal{H}_i}]:\gamma\in C_e^{\infty}(\mathbb{R},G)\} \quad \text{are defined by}$$

$$V_i(\gamma)\;\mathrm{Exp}\,[\Pi_i(\gamma')v_{\mathcal{H}_i}] := \mathrm{Exp}\,[\Pi_i(\gamma\gamma')v_{\mathcal{H}_i}]$$

and extending by linearity.

As cyclic vectors we are going to use

$$v_i' := \exp\left[-\tfrac{1}{2}\,\|v_{\mathfrak{X}_i}\|^2\right]\,\mathrm{Exp}\,v_{\mathfrak{X}_i}\ .$$

(These are obviously unit vectors!)

As expectation values we obtain:

$$E(V^i(\gamma)) = \exp \int_{I_i} \langle \Pi(\gamma(x))v - v, v\rangle dx\ .$$

We now construct the von Neumann tensor product of the L_i's (for further details see e.g. [11]) and denote this by $\bigotimes_{i\in\mathbb{N}} L_i$ with respect to the reference vector $(v_i')_{i\in\mathbb{N}_i}$. We can also construct the tensor product of the representations V^i, namely $V' := \bigotimes_{i\in\mathbb{N}} V^i$ which acts in $\bigotimes_{i\in\mathbb{N}} L_i$ (cf. [11], p. 153).

V' is then also a representation of $C_e^\infty(\mathbb{R},G)$ which is cyclic with respect to the subspace generated by the reference vector $\bigotimes_{i\in\mathbb{N}} v_i'$. As expectation value for V' we then obtain:

$$\begin{aligned}
E(V'(\gamma)) &= \langle \bigotimes_{i\in\mathbb{N}} V^i(\gamma)\,(\bigotimes_{i\in\mathbb{N}} v_i'),\,(\bigotimes_{i\in\mathbb{N}} v_i')\rangle \\
&= \prod_{i=1}^{\infty} \langle V^i(\gamma)v_i', v_i'\rangle \\
&= \prod_{i=1}^{\infty} \exp \int_{I_i} \langle \Pi(\gamma(x))v - v, v\rangle dx \\
&= \exp \int_{\mathbb{R}} \langle \Pi(\gamma(x))v - v, v\rangle dx\ .
\end{aligned}$$

Thus this representation is equivalent again to our original representation. Since all the representations V^i are reducible it is clear that our representation must be reducible also. So in this case coboundaries also give only reducible representations.

7. Factorizable Representations and CTPs

In section 4 we constructed a CTP of projective representations. We recall that we obtained there a cyclic projective representation $(\tilde{U},\tilde{\sigma},\tilde{e})$ with

$$\tilde{\sigma}(\gamma_1,\gamma_2) =$$

$$= \exp i\{\int_{\mathbb{R}}[a(\gamma_1(x))+a(\gamma_2(x))-a(\gamma_1\gamma_2(x))+\mathrm{Im}<\delta(\gamma_2(x)),\delta(\gamma_1(x)^{-1})>]dx$$

$$<U(\gamma(x))\tilde{e},\tilde{e}> = \exp\{i\int_{\mathbb{R}}[a(\gamma(x))-\frac{1}{2}<\delta(\gamma(x)),\delta(\gamma(x))>]dx\} .$$

In section 5 we then constructed a cyclic projective representation $(U,\sigma,\exp\Delta(\tilde{e}))$ with

$$\sigma(\gamma_1,\gamma_2) = \exp i\ \mathrm{Im}\int_{\mathbb{R}}<\delta(\gamma_2(x)),\delta(\gamma_1(x)^{-1})>dx$$

$$<U(\gamma)\ \mathrm{Exp}\ \Delta(\tilde{e}),\mathrm{Exp}\ \Delta(\tilde{e})> = \exp\{-\frac{1}{2}\int_{\mathbb{R}}<\delta(\gamma(x)),\delta(\gamma(x))>dx\} .$$

Given U we can thus define a new cyclic projective representation $(W,\sigma',\mathrm{Exp}\ \Delta(\tilde{e}))$ by setting

$$W(\gamma) := \exp\{i\int_{\mathbb{R}} a(\gamma(x))dm(x)\}\ U(\gamma) .$$

Then W has the following properties:

(i) $W(\gamma_1)W(\gamma_2) = \sigma'(\gamma_1,\gamma_2)W(\gamma_1\gamma_2)$ where

$$\sigma'(\gamma_1,\gamma_2) \equiv \tilde{\sigma}(\gamma_1,\gamma_2)$$

(ii) $<W(\gamma)\ \mathrm{Exp}\ \Delta(\tilde{e}),\ \mathrm{Exp}\ \Delta(\tilde{e})> = <\tilde{U}(\gamma)\tilde{e},\tilde{e}>$

(iii) W is projectively equivalent to U .

From (ii) and (i) it is clear that W is unitarily equivalent to $\tilde{U}$. We have thus embedded our CTP of projective representations in Fock space via a projective equivalence. This is a very general and explicit version of the well-known Araki-Woods Embedding Theorem, cf. [21].

It should now be clear that further progress in our discussion can only be made if we are able to compute the first order cohomology group in question. A good deal of progress has been made in this direction, cf. e.g. [7], [8], [23]. The general problem, however, is still unsolved. In the next chapter we shall investigate this problem for so-called regular semi-direct products. We shall obtain some quite explicit results which will provide a fair number of interesting examples.

III. FIRST ORDER COHOMOLOGY GROUPS FOR CERTAIN SEMI-DIRECT PRODUCTS

1. The General Theory

Let H be a locally compact, separable group and let N be a locally compact, separable abelian group. Let further $h \mapsto \alpha_h$ be a homomorphism from H into the group of all automorphisms of N (this homomorphism has to have "suitable" topological properties as well)! Then we define:

(1.1) Definition:

The *semi-direct product* of H and N is given by:

$$H \circledS N := \{(h,n) : (h,n) \in H \times N\} \quad \text{as a set.}$$

The group operation is then described by:

$$(h_1,n_1)\cdot(h_2,n_2) := (h_1h_2, n_1\alpha_{h_1}(n_2)) \qquad \forall\ (h_i,n_i) \in H \times N .$$

If the homomorphism $h \mapsto \alpha_h$ possesses "suitable" topological properties, then $G := H \circledS N$ is a separable, locally compact group with respect to the product topology (cf. [15]).

Let now $\hat{N}$ denote the character group of N. Furnished with the compact-open topology this is also a separable, locally compact group (cf. [25]).

The H-action on N (by means of the homomorphism $h \mapsto \alpha_h$) now induces an H-action on $\hat{N}$ as follows:

$$(h\chi)(n) := \chi(\alpha_{h^{-1}}(n)) \qquad \forall\ (h,n,\chi) \in H \times N \times \hat{N} .$$

For all $\chi \in \hat{N}$ we set, $H\chi := \{h\chi : h \in H\}$, the *orbit* generated by χ and the H-action described above.

Further let $H_\chi := \{h : h \in H \text{ and } h\chi = \chi\}$ denote the *stability subgroup* of χ.

We also note that the H-action on $\hat{N}$ induces a $G=H \circledS N$ - action on $\hat{N}$ by means of

$$(h,n)(\chi) := h\chi \quad \forall\ (h,n,\chi) \in H \times N \times \hat{N} .$$

Finally we set $G_\chi := H_\chi \circledS N$ and obtain a theorem which is due to Mackey ([15]):

(1.2) Theorem:

Let a χ from each orbit in $\hat{N}$ be chosen and an irreducible representation L of H. We suppose that it is possible to choose a Borel set in $\hat{N}$ which meets every orbit in exactly one point. Then we have the following:

(i) $(h,n) \mapsto \chi(n)L_h$ is an irreducible representation of $G_\chi = H_\chi \circledS N$.

(ii) All irreducible representations of $H \circledS N$ are unitarily equivalent to representations, which are induced from representations of G_χ having the form described in (i).

Remark:

A semi-direct product with the property that a Borel set C in $\hat{N}$ as described in (1.2) can be chosen is called a *regular semi-direct product*. We are going to consider only regular semi-direct products in the sequel. We assert at this point that there is a sufficiently large number of regular semi-direct products to make the theory interesting. (We are going to deal with the problem in more detail when we come to consider concrete examples.)
Now let $\Pi: G \to G/G_\chi$ be the natural projection. Then there exists a bijection (which can be shown to preserve Borel sets) from $G/G_\chi \to G\chi$ given by

$$gG_\chi \mapsto g\chi$$

(Here $G\chi$ denotes the orbit of χ in $\hat{N}$ under the G-action!)

Using this identification we shall now give the explicit form of the induced representations; first of all we'll have to introduce some terminology:

Let μ be a measure on $G\chi$ and $\mu^h(E):=\mu(hE)$ for each Borel set $E \subsetneq G\chi$ and each $h \in H$. Then μ is called *quasi-invariant* if μ^h is equivalent to μ for all $h \in H$. If μ is quasi-invariant then the existence of the Radon-Nykodym derivative with respect to μ^h is, of course, guaranteed. Further let $\mathcal{H}$ be the Hilbert space in which the representation L (as in (1.2)) acts. We consider functions $f: G\chi \to \mathcal{H}$ with the property

$$\int_{G\chi} \langle f(\chi'), f(\chi')\rangle d\mu(\chi') < \infty .$$

This determines a Hilbert space with scalar product

$$\langle f_1, f_2\rangle := \int_{G\chi} \langle f_1(\chi'), f_2(\chi')\rangle d\mu(\chi') .$$

Finally we need a theorem of Kuratowski (see [19]) which guarantees the existence of a Borel cross-section $\rho: G\chi \to G$ with $\Pi \circ \rho = 1_{G\chi}$. We then obtain:

(1.3) Theorem:

Let $G=H \circledS N$ be a regular semi-direct product. Then the irreducible unitary representations are (up to unitary equivalence) described as follows:

$$(U_{(h,n)} f)(\chi') := \chi'(n)\lambda(h,h^{-1}\chi')^{\frac{1}{2}} C(h,h^{-1}\chi') f(h^{-1}\chi') .$$

The symbols have the following meaning:

$$\lambda(h,\chi') := \frac{d\mu}{d\mu^h}(\chi')$$

$$(h,\chi') \in H \times \hat{N}$$

$$C(h,\chi') := L_{\rho(h\chi')^{-1} h \rho(\chi')} \qquad (L \text{ irreducible representation of } H)$$

$$f \in \mathcal{L}^2(G\chi, \mathcal{H}, \mu) .$$

We now turn to the computation of the first order cocycles associated with the representations described in (1.3). For this it is necessary to analyze the cocycle identity

$$U_{(h_1,n_1)}\delta(h_2,n_2) = \delta((h_1,n_1)\cdot(h_2,n_2))-\delta(h_1,n_1) \qquad (*)$$

in detail.

First of all we show

(1.4) Lemma:

If the induced representation is trivial when restricted to N, then we have $U_{(h,n)} \equiv V_h$ where $h \mapsto V_h$ is an irreducible representation of H. If V acts in a Hilbert space $\mathcal{K}$, then each cocycle associated with U is of the form:

$$\delta(h,n) = \delta_1(h)+\eta(n) \quad \text{where}$$

(i) δ_1 is a cocycle associated with V.

(ii) $\eta : N \to \mathcal{K}$ is a continuous, additive homomorphism.

(iii) $V_h\eta(n) = \eta(\alpha_h(n)) \quad \forall\ (n,n) \in H \times N$.

Proof:
First of all we note that trivially

$$(h,n) = (e,n)\cdot(h,0)$$

(where 0 denotes the neutral element in N).
From this we obtain using (*)

$$\delta(h,n) = \delta(h,0)+\delta(e,n) \qquad \forall\ (h,n) \in H \times N .$$

We set $\delta_1(h) := \delta(h,0)$
$\eta(n) := \delta(e,n)$.

Then it follows again from the cocycle identity that

$$V_{h_1}\delta_1(h_2) = \delta_1(h_1h_2)-\delta_1(h_1) \quad \forall\, h_1,h_2 \in H \quad \text{and}$$

$$\eta(n_2) = \eta(n_1+n_2)-\eta(n_1) \ .$$

(i) and (ii) above are an immediate consequence;

(iii) follows from

$$U_{(h,O)}\delta(e,n) = \delta(h,\alpha_h(n))-\delta(h,O)$$

$$\delta(e,\alpha_h(n)) \ . \qquad \text{q.e.d.}$$

For a complete classification we need the following

(1.5) Lemma:

Let the induced representation now be nontrivial when restricted to N. In that case we obtain as associated cocycles only coboundaries.

Proof:
For brevity we set $D(h,\chi'):=\lambda(h,\chi')^{\frac{1}{2}}\,C(h,\chi')$ (the notation being the same as in (1.3)); then we obtain immediately

$$(U_{(h,n)}f)(\chi') = \chi'(n)D(h,h^{-1}\chi')f(h^{-1}\chi') \ .$$

From (*) above we now obtain for each cocycle δ associated with U :

$$\delta(h,n)(\chi') = \chi'(n)\delta(h,O)(\chi')+\chi'(n)f_1(\chi')-f_1(\chi')$$

$$\forall\ (h,n) \in H \times N, \text{ a.e.}\chi' \ .$$

Here f_1 is a fixed element from $\mathcal{L}^2(G\chi,\mathcal{H},\mu)$. (This follows since $U_{(h,n)}$, when restricted to N is by assumption nontrivial and thus the corresponding part of the cocycle δ must be a coboundary; cf. [20] for the classification of the cohomology in the case of abelian groups.) Substitution in the cocycle identity gives for the left-hand side:

$$U_{(h_1,n_1)}\delta(h_2,n_2)(\chi') =$$

$$\chi'(n_1)D(h_1,h_1^{-1}\chi')[(h_1^{-1}\chi')(n_2)\delta(h_2,0)(h_1^{-1}\chi')$$

$$+(h_1^{-1}\chi')(n_2)f_1(h_1^{-1}\chi')-f_1(h_1^{-1}\chi')] =$$

$$\chi'(n_1+\alpha_{h_1}(n_2))D(h_1,h_1^{-1}\chi')[\delta(h_2,0)(h_1^{-1}\chi')+f_1(h_1^{-1}\chi')]$$

$$-\chi'(n_1)D(h_1,h_1^{-1}\chi')f(h_1^{-1}\chi') \ .$$

We now utilize the fact that $\delta(h,0)$ is a cocycle associated with the induced representation restricted to H and thus obtain for the above expression:

$$U_{(h_1,n_1)}\delta(h_2,n_2)(\chi') =$$

$$\chi'(n_1+\alpha_{h_1}(n_2))[\delta(h_1h_2,0)(\chi')-\delta(h_1,0)(\chi')+$$

$$D(h_1,h_1^{-1}\chi')f_1(h_1^{-1}\chi')]-\chi'(h_1)D(h_1,h_1^{-1}\chi')f_1(h_1^{-1}\chi') \tag{1}$$

We now consider the right-hand side of the cocycle identity (*) and obtain:

$$\delta(h_1h_2,n_1+\alpha_{h_1}(n_2))(\chi')-\delta(h_1,n_1)(\chi') =$$

$$\chi'(n_1+\alpha_{h_1}(n_2))[\delta(h_1h_2,0)(\chi')+f_1(\chi')]-\chi'(n_1)\delta(h_1,0)(\chi')$$

$$-\chi'(n_1)f_1(\chi') \ . \tag{2}$$

Comparison of (1) and (2) now gives:

$$[\chi'(\alpha_{h_1}(n_2))-1]\delta(h_1,0)(\chi') =$$

$$[\chi'(\alpha_{h_1}(n_2))-1][D(h_1,h_1^{-1}\chi')f_1(h_1^{-1}\chi')-f_1(\chi')]$$

$$\forall\ (h_1,n_2)\in H\times N \quad \text{a.e.}\chi' \ .$$

Since the representation is by assumption nontrivial on N we obtain:

$$\delta(h,n)(\chi') = \chi'(n)D(h,h^{-1}\chi')f_1(h^{-1}\chi')-f_1(\chi') \quad \text{a.e.}\chi' \ .$$

Thus in this case δ is a coboundary and the proof of the lemma is complete. q.e.d.

From (1.4) and (1.5) we obtain immediately

(1.6) Theorem:

Let $G=H \circledS N$ be a regular semi-direct product. Then only non-trivial cocycles of the form

$$\delta(h,n) = \delta_1(h)+\eta(n)$$

are possible. These are associated with irreducible representations $(h,n) \mapsto V_h$; δ, η, V_h are as in (1.4).

Remark:

It is, of course, possible that there are no non-trivial cocycles at all.

2. The Cohomology of the Euclidean Motion Groups

In this section we shall, as an application of the theory given in section 1, describe the cohomology of the Euclidean Motion Groups. We shall thus consider $H \circledS N$ with $H=SO(n)$ and $N=\mathbb{R}^n$ where the $SO(n)$-action on $\mathbb{R}^n$ is just the natural one (i.e. $SO(n)$ acts as group of rotations). It turns out that we obtain the same $SO(n)$-action on $\hat{\mathbb{R}}^n \cong \mathbb{R}^n$.

(2.1) Definition:

The groups $SO(n) \circledS \mathbb{R}^n$ (with the natural $SO(n)$-action) will be called the *Euclidean Motion Groups*.

As orbits in $\hat{\mathbb{R}}^n$ we obtain spheres centred on the origin. It follows immediately that all the semi-direct products are regular and thus the theory described above is applicable.

There are indeed irreducible representations which are trivial on $\mathbb{R}^n$. Since $SO(n)$ is compact for every $n \in \mathbb{N}$ these representations must be finite dimensional. We obtain the following

(2.2) Theorem:

The non-trivial cocycles of the groups $SO(n) \circledS \mathbb{R}^n$ are described by

$$\delta(A,\underline{x}) := C\underline{x},\ C \in \mathbb{R}, (A,\underline{x}) \in SO(n) \circledS \mathbb{R}^n .$$

They are associated with the representations

$$U_{(A,\underline{x})} := A \quad \text{in } \mathbb{R}^n .$$

Proof:
By (1.6) non-trivial cocycles have the form

$$\delta(A,\underline{x}) = \delta_1(A) + \eta(\underline{x}) . \qquad \text{(if they exist at all!)}$$

The associated representations have the form

$$U_{(A,\underline{x})} := V_A$$

where $A \mapsto V_A$ is an irreducible representation of $SO(n)$.

Since $SO(n)$ is compact and δ_1 is a cocycle associated with V (considered as representation of $SO(n)$) we have that δ_1 must be a coboundary (cf. [20]).

Thus we have to investigate the existence of a nontrivial, continuous, additive homomorphism $\eta: \mathbb{R}^n \to \mathcal{H}$ (where $\mathcal{H}$ is the Hilbert space in which V acts) which satisfies:

$$V_A \eta(\underline{x}) = \eta(A\underline{x}) \qquad (A,\underline{x}) \in SO(n) \circledS \mathbb{R}^n .$$

First of all we note that η must be linear since it is additive and continuous. We further note that $A \mapsto A$ is an irreducible representation of $SO(n)$ in $\mathbb{R}^n$. Thus if $\eta \neq 0$ we have a nontrivial intertwining operator for A (considered as representation) and V_A . Then we must have

$$OAO^{-1} = V_A$$

for some isometry O . From Schur's lemma it follows immediately that $\eta \equiv cO$ for some $c \in \mathbb{R}$. Since we are interested in representations only up to equivalence the statement of the theorem follows. q.e.d.

Remark:

The cocycles described in (2.2) are the so-called *Maurer-Cartan-cocycles* (cf. [22]). Thus other cocycles do not exist for the Euclidean Motion Groups.

3. The Cohomology of the First Leibniz-Extension of Compact Lie-Groups

In [22] the Leibnitz-Extension was defined in order to determine certain factorizable representations. It is thus of some interest to compute the relevant cohomology groups. So let G be a compact Lie group with Lie Algebra $\mathfrak{g}$. We consider G_L (cf. I.).

The crucial point is that G_L is a regular semi-direct product (cf. [14] p. 71). Thus again the theory of section 1 is applicable and we obtain:

(3.1) Theorem:

The first Leibnitz-Extensions of compact Lie groups have as nontrivial cocycles exactly the Maurer-Cartan cocycles.

Proof:

The proof is analogous to the proof of (2.2) above. Again we have to investigate the existence of a nontrivial, continuous, additive homomorphism $\eta: G \to \mathcal{H}$. This homomorphism now has to satisfy the condition

$$V_g \eta(X) = \eta(\mathrm{Ad}\, g(X)) \quad \forall\ (g,X) \in G \times \mathcal{H}$$

where $g \mapsto V_g$ is an irreducible representation of G in $\mathcal{H}$. Now η is $\mathbb{R}$-linear again and an application of Schur's lemma completes the proof as in (2.2). q.e.d.

We wish to deal rather more explicitly with a special case since it turns out that there is a connection between the first Leibnitz-Extension and a Euclidean Motion Group.

(3.2) The First Leibnitz-Extension of SO(3):

The Lie-Algebra of SO(3) is given by

$$so(3) := \left\{ \begin{bmatrix} 0 & a & b \\ -a & 0 & c \\ -b & -c & 0 \end{bmatrix} : a,b,c \in \mathbb{R} \right\}.$$

It is frequently not easy to give the explicit G-action on $\mathcal{G}$. A choice of a suitable basis makes things somewhat less complicated. Thus as a basis for so(3) we choose

$$X_1 = \begin{bmatrix} 0 & 0 & 0 \\ 0 & 0 & -1 \\ 0 & 1 & 0 \end{bmatrix},\ X_2 = \begin{bmatrix} 0 & 0 & 1 \\ 0 & 0 & 0 \\ -1 & 0 & 0 \end{bmatrix},\ X_3 = \begin{bmatrix} 0 & -1 & 0 \\ 1 & 0 & 0 \\ 0 & 0 & 0 \end{bmatrix}.$$

If now $\mathrm{ad}\, X: \mathcal{G} \to \mathcal{G}$ is defined by

$$\mathrm{ad}\, X: Y \mapsto [X,Y]$$

then we obtain as matrix representation $M(\mathrm{ad}\ X_i)$ with respect to the basis given above:

$$M(\mathrm{ad}\ X_i) = X_i \quad \text{for} \quad 1 \le i \le 3 \ .$$

We now note that each matrix $A \in SO(3)$ may be written as follows:

$$A = R(\alpha)\ S(\beta)\ R(\gamma) \quad \text{with}$$

$$R(\alpha) := \begin{pmatrix} \cos\alpha & -\sin\alpha & 0 \\ \sin\alpha & \cos\alpha & 0 \\ 0 & 0 & 1 \end{pmatrix} \qquad S(\beta) := \begin{pmatrix} \cos\beta & 0 & -\sin\beta \\ 0 & 1 & 0 \\ \sin\beta & 0 & \cos\beta \end{pmatrix} \ .$$

Thus to determine the group action in $\mathcal{G}$ completely it is only necessary to give the action of $R(\alpha)$ and $S(\beta)$. For this we make use of the relation

$$\mathrm{Ad}(\exp X) = e^{\mathrm{ad}\ X} \qquad \text{(cf. [12] p. 118)}$$

as well as of the fact that $\exp \alpha X_3 = R(\alpha)$ and $\exp \beta X_2 = S(-\beta)$.

Identifying $so(3)$ as a vector space with $\mathbb{R}^3$ by $X_i \mapsto e_i$, where $\{e_1, e_2, e_3\}$ is the standard basis in $\mathbb{R}^3$, we find that the $SO(3)$-action in $so(3)$ is simply the natural $SO(3)$-action in $\mathbb{R}^3$. Thus the first Leibnitz-Extension of $SO(3)$ may be considered as the Euclidean Motion Group $SO(3) \circledS \mathbb{R}^3$. The corresponding cohomology group is, of course, determined by the preceding theorem and indeed was used already in [9].

4. The First Leibniz-Extension of $SL(2;\mathbb{R})$

Here, of course, the group in question is not compact. This poses certain technical difficulties which will become apparent later on. The main obstacle to be overcome is to show that we are indeed dealing with a regular semi-direct product. (A general theorem, as in the case of compact groups, doesn't seem to be available here.)

(4.1) Definition:

$$SL(2;\mathbb{R}) := \{\begin{pmatrix} a & b \\ c & d \end{pmatrix} : ad-bc=1,\ a,b,c,d \in \mathbb{R}\} .$$

The group operation is the obvious one namely matrix multiplication.

The Lie algebra $sl(2;\mathbb{R})$ is then given by

$$sl(2;\mathbb{R}) := \{\begin{pmatrix} a & b \\ c & -a \end{pmatrix} : a,b,c \in \mathbb{R}\} .$$

The Lie bracket is the usual one.

To describe the Leibnitz-Extension explicitly we look for a suitable basis for $sl(2;\mathbb{R})$. A useful hint is contained in the following

(4.2) Lemma ([12] p. 223):

Let $\mathcal{G}$ be a semi-simple Lie algebra over $\mathbb{C}$. Then there exists a basis (X_i) for $\mathcal{G}$, such that the matrix representations $M(\mathrm{ad}\ X)$, $X \in \mathcal{G}$, for this basis possess the following properties (recall that since $\mathcal{G}$ is semi-simple we have the Iwasawa decomposition):

(i) $M(\mathrm{ad}\ u)$ is skew-hermitian if u belongs to the compact part of $\mathcal{G}$.

(ii) $M(\mathrm{ad}\ n)$ is triangular if n belongs to the nilpotent part of $\mathcal{G}$.

(iii) $M(\mathrm{ad}\ h)$ is diagonal, with real entries on the diagonal for certain h belonging to the abelian part of $\mathcal{G}$.

Obviously (4.2) can only give a hint since it refers to complex Lie algebras. We consider, however, $SL(2;\mathbb{C})$ in place of $SL(2;\mathbb{R})$.

A computation of the basis according to (4.2) (for details of the construction see the cited reference) gives:

$$X_1 = \begin{bmatrix} 0 & 0 \\ \frac{1}{2} & 0 \end{bmatrix}, \quad X_2 = \begin{bmatrix} \frac{1}{2\sqrt{2}} & 0 \\ 0 & \frac{-1}{2\sqrt{2}} \end{bmatrix}, \quad X_3 = \begin{bmatrix} 0 & \frac{1}{2} \\ 0 & 0 \end{bmatrix} .$$

This gives a basis for $sl(2;\mathbb{R})$ also, which proves extremely useful in computations.

Let now $\alpha_1, \alpha_2, \alpha_3 \in \mathbf{sl}(2;\mathbb{R})$ be defined by:

$$\alpha_1 = \begin{bmatrix} 0 & x \\ -x & 0 \end{bmatrix}, \quad \alpha_2 = \begin{bmatrix} \lambda & 0 \\ 0 & -\lambda \end{bmatrix}, \quad \alpha_3 = \begin{bmatrix} 0 & \mu \\ 0 & 0 \end{bmatrix}$$

where $x, \lambda, \mu \in \mathbb{R}$.

Then, using the Iwasawa decomposition, we see immediately that every element in $SL(2;\mathbb{R})$ may be written as $\exp \alpha_1 \exp \alpha_2 \exp \alpha_3$. So we are able to describe in detail the $SL(2;\mathbb{R})$-action in $sl(2;\mathbb{R})$. First of all, however, we identify $sl(2;\mathbb{R})$ with $\mathbb{R}^3$ as a vector space by setting $X_i \to e_i$, $1 \le i \le 3$, where $\{e_1, e_2, e_3\}$ is again the standard basis of $\mathbb{R}^3$. Then a lengthy (but straightforward) calculation gives the following results:

Let $M(\text{ad } \alpha_i)$ be the matrix representations of the $\text{ad } \alpha_i$ with respect to $\{X_1, X_2, X_3\}$ or $\{e_1, e_2, e_3\}$. Then we have:

$$\exp M(\text{ad } \alpha_1) \begin{bmatrix} X_1 \\ X_2 \\ X_3 \end{bmatrix} = \begin{bmatrix} \frac{1}{2}(\cos 2x+1) & \frac{-\sin 2x}{2} & \frac{1}{2}(\cos 2x-1) \\ \frac{\sin 2x}{2} & \cos 2x & \frac{\sin 2x}{2} \\ \frac{1}{2}(\cos 2x-1) & \frac{\sin 2x}{2} & \frac{1}{2}(\cos 2x+1) \end{bmatrix} \begin{bmatrix} X_1 \\ X_2 \\ X_3 \end{bmatrix}$$

$$\exp M(\text{ad } \alpha_2) \begin{bmatrix} X_1 \\ X_2 \\ X_3 \end{bmatrix} = \begin{bmatrix} e^{-2\lambda} & 0 & 0 \\ 0 & 1 & 0 \\ 0 & 0 & e^{2\lambda} \end{bmatrix} \begin{bmatrix} X_1 \\ X_2 \\ X_3 \end{bmatrix}$$

$$\exp M(\text{ad } \alpha_3) \begin{bmatrix} X_1 \\ X_2 \\ X_3 \end{bmatrix} = \begin{bmatrix} 1 & 0 & 0 \\ \sqrt{2}\mu & 1 & 0 \\ -\mu^2 & -\sqrt{2}\mu & 1 \end{bmatrix} \begin{bmatrix} X_1 \\ X_2 \\ X_3 \end{bmatrix}$$

Thus the $SL(2;\mathbb{R})$-action on $sl(2;\mathbb{R})$ or $\mathbb{R}^3$ has been completely described.

Let B denote the Cartan-Killing form for $sl(2;\mathbb{R})$. Then B is non-degenerate since $sl(2;\mathbb{R})$ is semi-simple. Thus we get a canonical (i.e. independent of the basis) identification between $sl(2;\mathbb{R})$ and $\hat{sl}(2;\mathbb{R})$ (considered as $\mathbb{R}^3$ and $\hat{\mathbb{R}}^3$) using B :

$$\phi: \quad \underline{p} \in \mathbb{R}^3 \mapsto \chi_{\underline{p}} \in \hat{\mathbb{R}}^3 \qquad \text{where}$$

$$\chi_{\underline{p}}(\underline{q}) := e^{iB(\underline{p},\underline{q})} \qquad \underline{p},\underline{q} \in \mathbb{R}^3 .$$

We note that with this identification the $SL(2;\mathbb{R})$-action in $\hat{\mathbb{R}}^3$ is the same as the $SL(2;\mathbb{R})$-action in $\mathbb{R}^3$, for if $g \in SL(2;\mathbb{R})$, we obtain from the invariance of the Cartan-Killing form:

$$\begin{aligned}(g\,\chi\underline{p})(\underline{q}) &= \chi_{\underline{p}}(\mathrm{Ad}\, g^{-1}(\underline{q})) = e^{iB(\underline{p},\mathrm{Ad}\, g^{-1}(\underline{q}))} \\ &= e^{iB(\mathrm{Ad}\, g\underline{p},\underline{q})} \\ &= \chi_{\mathrm{Ad}\, g(\underline{p})}(\underline{q}) .\end{aligned}$$

Thus we obtain the following commutative diagram:

$$\begin{array}{ccc} \mathbb{R}^3 & \xrightarrow{\mathrm{Ad}} & \mathbb{R}^3 \\ \phi \downarrow & & \downarrow \phi \\ \hat{\mathbb{R}}^3 & \xrightarrow[\text{Action}]{\text{Induced}} & \hat{\mathbb{R}}^3 \end{array}$$

A straightforward (but again somewhat lengthy) computation shows that the orbits under the $SL(2;\mathbb{R})$-action in $\hat{\mathbb{R}}^3 \doteq \mathbb{R}^3$ are described as follows:

a) $$\left\{ \begin{pmatrix} x_1 \\ x_2 \\ x_3 \end{pmatrix} : x_2^2 + 2x_1x_3 = 0,\ x_1 < 0,\ x_3 \geq 0 \right\} \cup$$

$$\left\{ \begin{pmatrix} x_1 \\ x_2 \\ x_3 \end{pmatrix} : x_1 = x_2 = 0,\ x_3 > 0 \right\}$$

b) $\left\{ \begin{pmatrix} x_1 \\ x_2 \\ x_3 \end{pmatrix} : x_2^2+2x_1x_3 = 0,\ x_1>0,\ x_3\leq 0 \right\} \cup$

$\left\{ \begin{pmatrix} x_1 \\ x_2 \\ x_3 \end{pmatrix} : x_1 = x_2 = 0,\ x_3<0 \right\}$

c) $\left\{ \begin{pmatrix} x_1 \\ x_2 \\ x_3 \end{pmatrix} : x_2^2+2x_1x_3 = r_1,\ r_1\in \mathbb{R},\ r_1>0 \right\}$

d) $\left\{ \begin{pmatrix} x_1 \\ x_2 \\ x_3 \end{pmatrix} : x_2^2+2x_1x_3 = r_2,\ r_2\in \mathbb{R},\ r_2<0,\ x_1>0,\ x_3<0 \right\}$

e) $\left\{ \begin{pmatrix} x_1 \\ x_2 \\ x_3 \end{pmatrix} : x_2^2+2x_1x_3 = r_3,\ r_3\in \mathbb{R},\ r_3<0,\ x_1<0,\ x_3>0 \right\}$

f) $\left\{ \begin{pmatrix} 0 \\ 0 \\ 0 \end{pmatrix} \right\}$

We are now able to give a Borel set C which meets every orbit in exactly one point; we take C as

$$C := \left\{\begin{pmatrix} 0 \\ 0 \\ 1 \end{pmatrix}\right\} \cup \left\{\begin{pmatrix} 0 \\ 0 \\ -1 \end{pmatrix}\right\} \cup \left\{\begin{pmatrix} 0 \\ \sqrt{r_1} \\ 0 \end{pmatrix}\right\} \cup \left\{\begin{pmatrix} \sqrt{\frac{r_2}{2}} \\ 0 \\ -\sqrt{\frac{r_2}{2}} \end{pmatrix}\right\} \cup \left\{\begin{pmatrix} -\sqrt{\frac{r_3}{2}} \\ 0 \\ \sqrt{\frac{r_3}{2}} \end{pmatrix}\right\} \cup \left\{\begin{pmatrix} 0 \\ 0 \\ 0 \end{pmatrix}\right\}$$

Thus we have shown that the first Leibnitz-Extension is a regular semidirect product and we can proceed to apply the theory. The result is

(4.3) Theorem:

There are exactly two non trivial cocycles for the Leibnitz-Extension of $SL(2;\mathbb{R})$. These will be described in detail later since they are just the two non-trivial cocycles for SL(2;$\mathbb{R}$) .

Proof:

We recall that there are no finite-dimensional unitary representations for SL(2;$\mathbb{R}$) . It is then an almost immediate consequence, that a Maurer-Cartan cocycle η cannot exist in this case. q.e.d.

IV. FIRST ORDER COHOMOLOGY FOR SL(2; ℝ) AND SL(2; ℂ)

As we have seen in the chapters before it is important for our applications to know all the solutions of the cocycle equation especially those which are not coboundaries. In this chapter we are mainly concerned with solving this problem for SL(2; ℝ) . To this end we must first construct all irreducible unitary representations of SL(2; ℝ) . We do this using the inducing construction described in [20] inducing all the series from the same subgroup. The result is not only very convenient for our purposes but also gives a new realization of the discrete series. These are the contents of the first four sections. In the fifth section we construct all non-trivial cocycles of SL(2; ℝ) associated with irreducible unitary representations explicitly. By generalizing a theorem of Parthasarathy and Schmidt about cocycles of induced representations (see [20]) we obtain a formula for certain solutions of the cocycle equation. A further discussion of the analytic vectors then shows that we have already obtained all non-trivial cocycles by applying the formula.
As the arguments for SL(2;ℂ) are largely analoguous we only summarize the results described in [7]. This is done in the last section of this chapter.

1. Preliminaries

Let SL(2; ℝ) be the Lie group of all real (2x2)-matrices with determinant 1 . It will be convenient to use an isomorphic Lie group instead, namely

$$SU(1,1) = \left\{\begin{bmatrix} \alpha & \beta \\ \beta & \alpha \end{bmatrix} : \alpha , \beta \in \mathbb{C} , |\alpha|^2-|\beta|^2 = 1\right\}$$

A simple computation then gives us the Lie algebra $\mathfrak{g}$ of SU(1,1) :

(1.1) Lemma:

(i) $$\mathfrak{g} = \left\{\begin{bmatrix} ia & b+ic \\ b-ic & -ia \end{bmatrix} : a,b,c \in \mathbb{R} \right\}$$

(ii) A basis of $\mathfrak{g}$ (considered as an $\mathbb{R}$ vector space) is obviously given by:

$$A = \begin{bmatrix} i & 0 \\ 0 & -i \end{bmatrix} \qquad B = \begin{bmatrix} 0 & 1 \\ 1 & 0 \end{bmatrix} \qquad C = \begin{bmatrix} 0 & i \\ -i & 0 \end{bmatrix}$$

with the commutation relations

$$[A,B] = 2C \qquad [B,C] = -2A \qquad [C,A] = 2B \ .$$

As $SU(1,1)$ is semi-simple it has an Iwasawa decomposition.

(1.2) Definition:

(i) For $\theta, s, t \in \mathbb{R}$ we define

$$k(\theta) := \begin{bmatrix} e^{i\theta} & 0 \\ 0 & e^{-i\theta} \end{bmatrix} \qquad n(s) := \begin{bmatrix} 1+is & -is \\ is & 1-is \end{bmatrix}$$

$$a(t) := \begin{pmatrix} \cosh t & \sinh t \\ \sinh t & \cosh t \end{pmatrix}$$

(ii) $K := \{k(\theta) : \theta \in \mathbb{R}\}$; $N := \{n(s) : s \in \mathbb{R}\}$,
$A := \{a(t) : t \in \mathbb{R}\}$

are clearly one parameter subgroups of $SU(1,1)$, where K is compact, N nilpotent, and A abelian.

(1.3) Lemma: (Iwasawa decomposition)

Let $g = \begin{pmatrix} \alpha & \beta \\ \bar{\beta} & \bar{\alpha} \end{pmatrix} \in SU(1,1)$.

Then there exist uniquely determined $k \in K$, $n \in N$, and $a \in A$ with $g=kna$.

The parameters $\theta \in \mathbb{R}$, $s \in \mathbb{R}$, and $t \in \mathbb{R}$ (see (1.2)) are given by:

$t = \ln |\alpha + \beta|$

$\Theta = \arg (\alpha+\beta)$

$s = \mathrm{Im}\ \alpha\bar{\beta}$.

Obviously N is a normal subgroup of the group $N\cdot A$. The centre Z of $SU(1,1)$, which is given by

$$Z = \{ \begin{bmatrix} 1 & 0 \\ 0 & 1 \end{bmatrix} , \begin{bmatrix} -1 & 0 \\ 0 & -1 \end{bmatrix} \} ,$$

is contained in K .

We define another subgroup D of $SU(1,1)$ by $D:=N\cdot(A\cdot Z)$. This will be the subgroup we are going to use for our inducing construction. As N is normal D is the semidirect product of N and $A\cdot Z$ and we can write the group multiplication in D by real parameters (see (1.2), parametrize Z by $r \in \{-1,1\}$) in the following manner:

$$(s_1,t_1,r_1)\cdot(s_2,t_2,r_2) = (s_1+s_2e^{2t_1},t_1+t_2,r_1 r_2)$$

$$\forall\ s_1,s_2,t_1,t_2 \in \mathbb{R} \quad \forall\ r_1,r_2 \in \{-1,1\} .$$

A straightforward calculation then gives us the quotient space $SU(1,1) / D$.

(1.4) Lemma:

(i) $SU(1,1) / D$ can be identified with $S^1 = \{z \in \mathbb{C} : |z|=1\}$ in the following manner:
The element $z=e^{i\theta} \in S^1$ represents the coset of all $g \in SU(1,1)$ having one of the two complex square roots of $e^{i\theta}$ as the "compact" parameter.

(ii) The canonical projection $\Pi : SU(1,1) \longrightarrow SU(1,1) / D\ (\cong S^1)$ is given by:

$$\Pi \left(\begin{bmatrix} \alpha & \beta \\ \bar{\beta} & \bar{\alpha} \end{bmatrix} \right) = \frac{\alpha+\beta}{\bar{\alpha}+\bar{\beta}} .$$

(iii) The natural action of $SU(1,1)$ on $SU(1,1) / D\ (\cong S^1)$ is then given by:

$$z \mapsto g\cdot z = \frac{\alpha z+\beta}{\bar{\alpha}+\bar{\beta}z} \qquad \forall z \in S^1 , \quad \forall g \in SU(1,1) .$$

2. The Construction of the Principal Series for SU(1,1)

We want to induce the principal series by the construction described in [20]. As the closed subgroup we take D (as defined in section 1.).

First we need a one-one Borel map $\rho : S^1 \to SU(1,1)$ with $\Pi \circ \rho = id_{S^1}$ (such a cross section always exists due to a theorem of Kuratowski). For an arbitrary $x \in S^1$ let $r(x)$ be a fixed square root of x (take for example for every x the well-determined root with an argument from $[0, \pi[)$.
Then

$$\rho(x) := \begin{bmatrix} r(x) & 0 \\ 0 & r(x)^{-1} \end{bmatrix}$$

is obviously a Borel map (not continuous!) which satisfies the above conditions.

Next we need a quasi invariant measure on S^1 . We take the standard Lebesgue measure μ_1 . Since μ_1 and μ_1^g (defined by $\mu_1^g(E) := \mu_1(gE)$ for any Borel set E) are equivalent the Radon Nikodym derivative

$$\frac{d\mu_1}{d\mu_1^g} =: \lambda(g,x) \quad \text{exists for } g \in SU(1,1) \text{ and a. e. } x(\mu_1) .$$

A simple computation using (1.4) yields:

$$\lambda(g,x) = |\bar{\alpha}+\bar{\beta}x|^2 .$$

For a Hilbert space V with inner product $(\ldots,\ldots)$ we can define $L_2(\mu_1,V)$ to be the space of all weakly measurable functions $f: SU(1,1)/D \to V$ satisfying

$$\int (f(x),f(x))\,d\mu_1(x) < \infty$$

and equip it with the inner product

$$<f_1,f_2>: = \int (f_1(x),f_2(x))d\mu_1(x) \ .$$

Let $d \mapsto L_d$ be any unitary representation of D . Then for every $g \in SU(1,1)$, $x \in S^1$

$$C(g,x): = L_{(\rho(gx))^{-1}g\rho(x)}$$

is an isometry on V (note that $\rho(gx)^{-1}g\rho(x) \in D$!).

Thus we now obtain a unitary representation

$$(U_g^L f)(x): = \lambda(g,g^{-1}x)^{\frac{1}{2}}C(g,g^{-1}x)f(g^{-1}x) \qquad (*)$$

of $SU(1,1)$ on $L_2(\mu_1,V)$.

For our purposes we only take one-dimensional unitary representations for L , i.e. unitary characters on D .

Since D is a semidirect product it is clear that a character is necessarily trivial on N . So we can distinguish between two kinds of characters namely those which are identity on the centre and those which are not, precisely:
If $r \in \{-1,1\}$ and $s \in \mathbb{R}$ are the parameters of Z and A respectively (see (1.3)) one has the unitary characters

a) $\quad \chi_1^{i\mu}(d) = (e^t)^{i\mu}$

b) $\quad \chi_2^{i\mu}(d) = r(e^t)^{i\mu}$

with an arbitrary $\mu \in \mathbb{R}$.

For such a given character we now have to compute $C(g,g^{-1}x)$. According to the definition of the characters we need expressions for e^t and $e^{i\theta}$ for $d = \rho(x)^{-1}g\rho(g^{-1}x) \in D$.

Using (1.4) and the definition of ρ we get:

$$\rho(x)^{-1}g\rho(g^{-1}x) = \begin{bmatrix} A & B \\ \overline{B} & \overline{A} \end{bmatrix} \qquad \text{with}$$

$$A = \alpha(r(x))^{-1}r\left(\frac{\bar{\alpha}x-\beta}{\alpha-\bar{\beta}x}\right)$$

$$B = \beta(r(x))^{-1}(r(\frac{\bar{\alpha}x-\beta}{\alpha-\bar{\beta}x}))^{-1} .$$

The formulas given in (1.3) yield:

$$e^{t} = \frac{1}{|\alpha-\bar{\beta}x|} , \quad \text{and} \quad e^{i\theta} = (r(x))^{-1}(r(\frac{\bar{\alpha}x-\beta}{\alpha-\bar{\beta}x}))^{-1} \frac{x|\alpha-\bar{\beta}x|}{\alpha-\bar{\beta}x} .$$

With these expressions and the Radon Nikodym derivative $\lambda(g,x)$ explicit formulas for the principal series are obtained from (*) . Because of the form of $e^{i\theta}$, however, the representation induced from $\chi_2^{i\mu}$ is a little complicated. If we take instead of $U^{\chi_2^{i\mu}}$ the unitarily equivalent representation given by $CU^{\chi_2^{i\mu}}C^{-1}$, where $C:L_2(S^1)\longrightarrow L_2(S^1)$ is multiplication by $r(x)$, we get a more convenient form.

Summarizing our results we have

(2.1) Theorem: (Principal Series)

Let $q=i\tau$ (with $\tau \in \mathbb{R}$), let $L_2(S^1)$ be the space of all square integrable functions with period 2π (equipped with the canonical inner product).

Then

(i) $$(U_g^q f)(z) = \frac{\alpha-\bar{\beta}z}{|\alpha-\bar{\beta}z|^{q+1}} f(\frac{\bar{\alpha}z-\beta}{\alpha-\bar{\beta}z})$$

and

(ii) $$(V_g^q f)(Z) = \frac{\alpha-\bar{\beta}z}{|\alpha-\bar{\beta}z|^{q+2}} f(\frac{\bar{\alpha}z-\beta}{\alpha-\bar{\beta}z})$$

with $g = \begin{bmatrix} \alpha & \beta \\ \bar{\beta} & \bar{\alpha} \end{bmatrix} \in SU(1,1)$, $f \in L_2(S^1)$

are unitary representations of $SU(1,1)$ on $L_2(S^1)$, the so-called principal series, which are induced from unitary characters of D , namely

(i) $\chi_1^q(d) = (e^t)^q$

resp.

(ii) $\chi_2^q(d) = r(e^t)^q$.

(2.2) Remarks:

(i) We have not yet considered the question of irreducibility. As we shall see in the next section all these representations (except V^q with $q=0$) are irreducible.

(ii) For any $q \in \mathbb{C}$ χ_1^q and χ_2^q are (in general non-unitary) characters on D . So for any $q \in \mathbb{C}$ U^q and V^q are formally "representations" (homomorphisms) which are, of course, not unitary for the canonical inner product of $L_2(S^1)$. The question now arises if one can take a suitable subspace of $L_2(S^1)$ and furnish it with a new inner product structure so that U^q resp. V^q becomes a unitary representation. In the next section we want to find <u>necessary</u> conditions for $q \in \mathbb{C}$ to yield a unitary representation in the above manner.

3. <u>Necessary Conditions for the Unitarity of Induced Representations of SU(1,1)</u>

In this section we want to find out for which $q \in \mathbb{C}$ U^q or V^q (as given in theorem (2.1)) can be extended to a unitary representation, we defer the question if such representations really exist and especially what the inner products look like to the next section. Our arguments are largely similar to those given in S. Lang's book (see [13], Chapter VI), so we shall only sketch the main ideas.

Let Π be any irreducible unitary representation of SU(1,1) in a Hilbert space $\mathcal{H}$.

Let $\mathcal{H}_\pi^{an} \subseteq \mathcal{H}$ be the space of analytic vectors of Π .

The importance of the derived representation $d\Pi$ of the Lie algebra of SU(1,1) is stated in the next

(3.1) Theorem: (see [13], p. 99)

Let G be connected, let V be an (algebraic) subspace of $\mathcal{H}_{\Pi}^{an}$ which is invariant under $d\Pi(X)$ for any $X \in \mathfrak{g}$. Then the closure of V is a G-invariant subspace of $\mathcal{H}$.

From this theorem it is clear what we are going to do now. We shall consider $d\Pi(X)$ in order to find (minimal) invariant subspaces and take the closure thus getting irreducible representations of SU(1,1).

(3.2) Definition:

Let $n \in \mathbb{Z}$, let $K = \{k(\theta) \in SU(1,1); \theta \in \mathbb{R}\}$ be the compact subgroup in the Iwasawa decomposition (see (1.3)).

Then we define a subspace $H_n \subseteq \mathcal{H}$ by

$$H_n := \{v \in \mathcal{H} : \Pi_{k(\theta)} v = e^{in\theta} v \quad \forall \theta \in \mathbb{R}\} .$$

H_n is called the *n-th eigenspace of* K *in* $\mathcal{H}$.

Some important properties of the H_n's which can all be found in S. Lang's book (see [13], Chapter VI and X) are summarized in

(3.3) Theorem:

Let Π be a unitary representation of SU(1,1), K be the compact subgroup in the Iwasawa decomposition which is clearly generated by $A \in \mathfrak{g}$ (see (1.1)).

Then:

(i) dim $H_n=0$ or dim $H_n=1$ $\forall\ n \in \mathbb{Z}$

(ii) $n \neq m \Rightarrow H_n \perp H_m$ $\forall\ n,m \in \mathbb{Z}$

(iii) $H_n \subseteq \mathcal{H}_\Pi^{an}$ $\forall\ n \in \mathbb{Z}$

(iv) H_n = eigenspace of $d\Pi(A)$ for the eigenvalue in .

By (3.3)(iv) we already know the action of $d\Pi(A)$ on such an H_n . Thus we only have to study the action of the two other elements of $\mathfrak{g}$ on H_n , which together with A form a basis of $\mathfrak{g}$. It will be convenient therefore to take the complexification $\tilde{\mathfrak{g}}$ instead of the real Lie algebra $\mathfrak{g}$ and extend the derived representation to $\tilde{\mathfrak{g}}$ by

$$d\Pi(X+iY) := d\Pi(X)+id\Pi(Y) .$$

Consideration of the eigenvalue problem of the adjoint representation leads to the following choice of a basis of $\tilde{\mathfrak{g}}$:

If {A,B,C} are as in (1.1), {A,R,S} is obviously a basis of $\tilde{\mathfrak{g}}$, where

$$R := B-iC \quad \text{and} \quad S := B+iC .$$

Now we are able to study the action of $d\Pi(\tilde{\mathfrak{g}})$ on the H_n's by a straightforward computation using the commutation rules and (3.3)(iv):

(3.4) Lemma:

Let {A,R,S } be the basis of $\tilde{\mathfrak{g}}$ defined above. Then for any n holds:

$$d\Pi(A)H_n \subseteq H_n$$

$$d\Pi(R)H_n \subseteq H_{n+2}$$

$$d\Pi(S)H_n \subseteq H_{n-2} .$$

Especially $d\Pi(\tilde{\mathfrak{g}})$ leaves $\sum_{n \text{ even}} H_n$ as well as $\sum_{n \text{ odd}} H_n$ invariant. Consequently if Π is irreducible (and we are only interested in such representations) one of the two algebraic sums must be zero. We then say that Π is of even resp. odd parity.

But yet another thing can happen. It may be possible that there is an m with $d\Pi(R)H_m=\{0\}$ or $d\Pi(S)H_m=\{0\}$. We then call Π a representation of highest (resp. lowest) weight m.

It should be clear now that because of the irreducibility one can distinguish between the following cases:

$\mathcal{H}$ is the closure of

1. $\sum_{n \text{ even}} H_n$ resp. $\sum_{n \text{ odd}} H_n$ or

2. $\sum_{\substack{n \leq m \\ n \text{ even}}} H_n$ resp. $\sum_{\substack{n \leq m \\ n \text{ odd}}} H_n$ or

3. $\sum_{\substack{n \geq m \\ n \text{ even}}} H_n$ resp. $\sum_{\substack{n \geq m \\ n \text{ odd}}} H_n$.

Note that all the H_n's which occur in the sums are non-zero!

Until now we have mainly made use of the irreducibility and not of the unitarity of Π. Doing that we shall get a more detailed classification. Let us first consider the case where $\mathcal{H}$ is the closure of $\sum_{n \text{ even}} H_n$. We have already noted that each H_n occurring is one-dimensional, so we can choose and fix a vector $v_0 \in H_0$ with $\|v_0\|=1$.

The definition

$$v_{n+2} := d\Pi(R)v_n \quad \forall\, n \in \mathbb{Z}$$

then gives us an orthogonal basis for $\sum_{n \text{ even}} H_n$ by recursion. As $d\Pi(S)$ maps H_n into H_{n-2} we have implicitly defined a sequence of complex numbers c_n by $d\Pi(S)v_n = c_n v_{n-2}$.

Using the commutation relations a recursion formula for the c_n is obtained:

$$c_n - c_{n+2} = 4n \qquad (*) .$$

If we now make use of the fact that because of the unitarity of Π $d\Pi(X)$ is skew-symmetric for every $X \in \mathfrak{g}$ (not $X \in \tilde{\mathfrak{g}}$!) , we can prove that the c_n must be real and negative.

Due to (*) the whole sequence of the c_n's is well determined if one has chosen an arbitrary negative $c_0 \in \mathbb{R}$. It should be obvious that the same arguments go through if we consider $\sum_{n \text{ odd}} H_n$ or the other cases. Always the same recursion formula (*) is obtained. For $H = \sum_{n \text{ odd}} H_n$ we choose an arbitrary negative $c_1 \in \mathbb{R}$ for the definition of the whole sequence.

If Π has lowest weight m then $d\Pi(S)v_m = 0$ and we can similarly show that (*) holds for $n \geq m$ and $c_{m+2} = -4m$.

So we have already a fixed value for the beginning of our recursion and don't have the free choice of a real parameter. Another consequence is that (since the c_n are negative) a lowest weight must be positive.

In the same way in the highest weight case the sequence c_n is well-determined and a highest weight is necessarily negative.

Using the techniques of infinitesimal isomorphisms it isn't hard to prove that two given irreducible unitary representations are unitarily equivalent iff the corresponding sequences c_n are equal. Thus we obtain a complete classification of all irreducible unitary representations by choosing an arbitrary negative real c_0 resp. c_1 (without weight vectors) or by indicating the lowest (highest) weight taking into consideration that the weight must be a positive (negative) integer.

We are now able to answer the question for which $q \in \mathbb{C}$ U^q or V^q as given in (2.1) can describe a unitary representation (on a suitable pre-Hilbert space of functions $f: S^1 \to \mathbb{C}$). So let us assume that $U^q (V^q)$ is already an irreducible unitary representation. If we define (for convenience) H_n to be the n-th eigenspace of K for U^q and I_n that of V^q it is almost trivial that for each $n \in \mathbb{Z}$ $\psi_n \in H_{-2n}$ resp. $\psi_n \in I_{-2n+1}$ where we have set $\psi_n(z) = z^n$ (with $z \in S^1$).

So the $\{\psi_n : n \in \mathbb{Z}\}$ (or a subset of it) must form an orthogonal basis for the representation space. Another consequence is that U^q is of even and V^q of odd parity.

For a more detailed discussion we first have to analyse the action of $dU^q(\tilde{\mathfrak{g}})$ and $dV^q(\tilde{\mathfrak{g}})$ on the ψ_n's . A straight-forward (but lengthy) computation leads to:

(3.5) Lemma:

Let $\{A,R,S\}$ be the basis of $\tilde{\mathfrak{g}}$ defined above. Then:

(i) $dU^q(A)\ \psi_n = -2in\ \psi_n$

$dU^q(R)\ \psi_n = (q+1-2n)\ \psi_{n-1}$

$dU^q(S)\ \psi_n = (q+1+2n)\ \psi_{n+1}$

(ii) $dV^q(A)\ \psi_n = -i(2n-1)\ \psi_n$

$dV^q(R)\ \psi_n = (q+2-2n)\ \psi_{n-1}$

$dV^q(S)\ \psi_n = (q+2n)\ \psi_{n+1}$.

Let us now first consider the case of even parity without weight vectors. Then all these representations are determined (up to unitary equivalence) by the choice of a real negative constant c_0 .

As $[\psi_0]=H_0$ (note that all the occurring H_n's are one-dimensional), there exists an $\alpha \neq 0$ so that $v_0 = \alpha\psi_0$ is a unit vector. We can take this v_0 as initial vector for our recursion formula for the sequence v_n .

Then:

$$v_2 = dU^q(R)v_0 = \alpha dU^q(R)\ \psi_0 \underset{(3.5)}{=} \alpha(q+1)\ \psi_{-1}\ .$$

Using the definition of the c_n's we get:

$$c_2 v_0 = dU^q(S)v_2 = \alpha(q+1)dU^q(S)\ \psi_{-1} \underset{(3.5)}{=} \alpha(q+1)(q-1)\ \psi_0 = (q+1)(q-1)v_0\ .$$

Because of (*) we have $c_2=c_0$ and are lead to

$$c_0 = q^2-1 < 0 \Leftrightarrow q \in i\mathbb{R} \vee q \in]-1,1[\setminus\{0\} .$$

Another consequence is that U^q and $U^{\tilde{q}}$ are unitarily equivalent iff $q=\tilde{q}$ or $q=-\tilde{q}$. It should be emphasized that we have already constructed representations for $q \in i\cdot\mathbb{R}$, to do the same for $q \in]-1,1[\setminus\{0\}$ is the aim of the next section.

Analogously we get in the case of odd parity $c_1=q^2$, so $q \in i\cdot\mathbb{R}\setminus\{0\}$.

Let us now assume that U^q is a representation with lowest weight $m \in \mathbb{Z}$. Since U^q is of even parity, m must be even, and, as we have seen above, m is necessarily positive, say $m=2l$ with $l \in \mathbb{N}_+$.

Due to the definition of a lowest weight vector $dU^q(S)H_m$ must be zero. Because of $H_{2m}=H_{2l}=[\psi_{-l}]$ (3.5) yields:

$$0 = dU^q(S)\psi_{-l} = (q+1-2l)\psi_{-l+1}$$
$$\Rightarrow \quad q = 2l-1 .$$

As the representation space H_-^q we take the closure of $\sum_{\substack{n \geq m \\ n \text{ even}}} H_n$.

Consequently $\{\psi_k : k \in \mathbb{Z} \wedge k \leq -l\}$ must be an orthogonal basis of H_-^q.

If U^q has highest weight m, m must be even and negative, say $m=-2l$ with $l \in \mathbb{N}_+$. We again get $q=2l-1$, but this time $\{\psi_k : k \in \mathbb{Z} \wedge k \geq l\}$ must be an orthogonal basis for the representation space H_+^q.

So we have the situation that for a given odd positive q U^q can describe two irreducible representations on different subspaces H_-^q and H_+^q having only $\{0\}$ as their intersection, namely one with lowest weight $q+1$ on H_-^q and the other with highest weight $-q-1$ on H_-^q. In the same way we can consider the cases of weight vectors for V^q. The result of this discussion is summarized with the other results of this section in

(3.6) Theorem:

For $q \in \mathbb{C}$ let U^q and V^q algebraically be given as in (2.1). Then U^q resp. V^q can be extended to irreducible unitary representations only in the following cases:

(i) $q \in i\cdot\mathbb{R}$ for U^q and $q \in i\cdot\mathbb{R}\setminus\{0\}$ for V^q ; the representation space then must have $\{\psi_k : k \in \mathbb{Z}\}$ as an orthogonal basis;

(ii) $q \in]-1,1[\setminus\{0\}$ only for U^q , the representation space must have $\{\psi_k : k \in \mathbb{Z}\}$ as an orthogonal basis;

(iii) $q=2l-1 (l \in \mathbb{N}_+)$ for U^q , which is then of lowest weight $q+1$ on a space spanned by $\{\psi_k : k \in \mathbb{Z} \wedge k \leq -1\}$ or of highest weight $-q-1$ on a space spanned by $\{\psi_k : k \in \mathbb{Z} \wedge k \geq 1\}$;

(iv) $q=2l-2 (l \in \mathbb{N}_+)$ for V^q , which is then of lowest weight $q+1$ on a space spanned by $\{\psi_k : k \in \mathbb{Z} \wedge k \leq -l+1\}$ or of highest weight $-q-1$ on a space spanned by $\{\psi_k : k \in \mathbb{Z} \wedge k \geq l\}$.

Remark:

Until now we only have realisations for (i) and for $l=1$ in (iv), namely those given in (2.1). But by the considerations above it is already clear that all the representations of the principal series (except V^q with $q=0$) are irreducible. So we will no longer call V^q with $q=0$ a member of the principal series but of the discrete series. It will be the aim of the next section to construct inner products for the remaining cases.

4. The Complementary and the Discrete Series of SU(1,1)

We now want to construct inner products for those $q \in \mathbb{C}$ for which according to (3.6) unitarization is possible but for which we have not yet got an inner product. All those representations are induced from non-unitary characters. There is - to our knowledge - no standard method for doing this. Some rough ideas about induction from non-unitary representations of a subgroup can be obtained from Mackey's lecture notes (see [16]); for details of the adaptation to our description of induced representations see [4], 2.3 .

First we observe that $C_0(S^1)$, the space of all continuous complex-valued functions defined on S^1 , is obviously stable under the action of all the U^q's and V^q's . Following Mackey's ideas we can finally reduce the construction of an inner product to the investigation of the existence of an antilinear operator $T:C_0(S^1) \to C_0(S^1)$ (or a "suitable" subspace), which has the following properties:

If one defines a sesquilinear form $\langle \dots, \dots \rangle_T$ on $C_0(S^1)$ by

$$\langle f,g\rangle_T = \int_0^{2\pi} (Tg)(x) f(x) dx$$

$\langle \dots, \dots \rangle_T$ will become an inner product for which U^q resp. V^q can be extended to a unitary representation. As T is required to act on a space of continuous functions one tries to get the operator in the form

$$(Tg)(x) = \int_0^{2\pi} k(x,y)\bar{g}(y) dy \ .$$

Let us first consider the case described in (3.6) (ii). Since as a consequence of our former arguments U^q and U^{-q} must be unitarily equivalent, we restrict ourselves to $q \in \,]0,1[$.

We can now prove that for any $\sigma \in \,]-\infty,\frac{1}{2}[$ and $f \in C_0(S^1)$

$\theta \mapsto \int_0^{2\pi} \frac{1}{|e^{i\theta}-e^{i\theta'}|^{2\sigma}} \bar{f}(e^{i\theta'}) d\theta'$ is a well-defined continuous function with period 2π. So T defined by the above integral is an antilinear operator on $C_0(S^1)$. On the other hand a lengthy computation using fourier expansions yields:

$$\langle f_1, f_2 \rangle := \int_0^{2\pi}\int_0^{2\pi} \frac{f_1(e^{i\theta})\bar{f}_2(e^{i\theta'})}{|e^{i\theta}-e^{i\theta'}|^{2\sigma}} d\theta \, d\theta'$$

is an inner product on $C_0(S^1)$ for any $\sigma \in \,]0,\frac{1}{2}[$. It is not very difficult to see that $\{\psi_k : k \in \mathbb{Z}\}$ is an orthogonal basis for $C_0(S^1)$ with this inner product.

Finally a straightforward calculation shows that

$$\langle U_g^q f, U_g^q f\rangle = \langle f,f\rangle \text{ holds on } C_0(S^1)\text{ , iff } \sigma = \frac{1-q}{2}\text{ .}$$

Since q is from $]0,1[$, we always get $\sigma \in]0,\frac{1}{2}[$, so in any case the sesquilinear form $\langle \ldots,\ldots\rangle$ is an inner product. Thus we have proved

(4.1) Theorem: (Complementary Series)

Let $q \in]0,1[$ for $g \in SU(1,1)$ let

$$(U_g^q f)(z) = \frac{1}{|\alpha - \bar{\beta} z|^{q+1}} f\left(\frac{\bar{\alpha} z-\beta}{\alpha - \bar{\beta} z}\right)$$

on $C_0(S^1)$. Then U^q is an irreducible unitary representation for the inner product $\langle \ldots,\ldots\rangle_q$ given by

$$\langle f_1,f_2\rangle_q = \int_0^{2\pi}\int_0^{2\pi} \frac{f_1(e^{i\theta}) f_2(e^{i\theta'})}{|e^{i\theta}-e^{i\theta'}|^{1-q}}\, d\theta\, d\theta'\ .$$

Completion of $C_0(S^1)$ (w.r.t. $\langle\ldots,\ldots\rangle_q$) gives the representation space, for which $\{\psi_k : k \in \mathbb{Z}\}$ is an orthogonal basis. This is (up to a constant) the same realisation of the complementary series as the one given by R. Takahashi in [27].

For the construction of the representations with weight vectors we first make

(4.2) Definition:

Let $l \in \mathbb{N}$, let $C_0(S^1)$ be the space of all complex-valued continuous functions on S^1 . Then we define the following subspaces of $C_0(S^1)$:

$$C_+^l(S^1) := \{ f \in C_0(S^1) : \int_0^{2\pi} e^{-inx} f(e^{ix})\, dx = 0 \quad \forall\ n < l \ \}$$

$$C_-^l(S^1) := \{ f \in C_0(S^1) : \int_0^{2\pi} e^{-inx} f(e^{ix})\, dx = 0 \quad \forall\ n > -l\ \}\ .$$

Trying an inner product analoguous to the one for the complementary series cannot be successful since because of the unitarity σ must be $\frac{1-q}{2}$, but then we have (for $q \geq 1$) no longer an inner product. Tedious calculations give the following results:

(4.3) Theorem: (Discrete Series, Even Weights)

Let $l \in \mathbb{N}_+$, $q=2l-1$ for $g \in SU(1,1)$ let

$$(U_g^q f)(z) = \frac{1}{|\alpha-\bar\beta z|^{2l}} f\left(\frac{\bar\alpha z-\beta}{\alpha-\bar\beta z}\right)$$

on $C_+^1(S^1)$ (resp. $C_-^1(S^1)$).

Then U^q is an irreducible unitary representation with highest (lowest) weight $-q-1$ $(q+1)$ on H_+^q (H_-^q) with respect to $\langle\ldots,\ldots\rangle_q$, where we have defined:

$$\langle f_1,f_2\rangle_q := (-1)^l \int_0^{2\pi}\int_0^{2\pi} |e^{i\theta}-e^{i\theta'}|^{2(l-1)} \ln\frac{|e^{i\theta}-e^{i\theta'}|^2}{2} f_1(e^{i\theta})\bar f_2(e^{i\theta'})\,d\theta d\theta'$$

and H_+^q (H_-^q) as the completion of $C_+^1(S^1)$ $(C_-^1(S^1))$ with $\langle\ldots,\ldots\rangle_q$. $\{\Psi_k : k\geq 1\}$ $(\{\Psi_k : k \leq -1\})$ is an orthogonal basis for H_+^q (H_-^q) . These representations belong to the discrete series, namely those with even weight.

(4.4) Theorem: (Discrete Series, Odd Weights Greater than 1)

Let $l \in \mathbb{N}$, $l\geq 2$, let $q=2l-2$. For $g \in SU(1,1)$ let

$$(V_g^q f)(z) = \frac{\alpha-\bar\beta z}{|\alpha-\bar\beta z|^{2l}} f\left(\frac{\bar\alpha z-\beta}{\alpha-\bar\beta z}\right)$$

on $C_+^l(S^1)$ (resp. $C_-^{l-1}(S^1)$).

Then V^q is an irreducible unitary representation with highest (lowest) weight $-q-1$ $(q+1)$ on H_+^q (H_-^q) with respect to $\langle\ldots,\ldots\rangle_q^+$ $(\langle\ldots,\ldots\rangle_q^-)$, where we have defined:

$$\langle f_1,f_2\rangle_q^+ := (-1)^l \int_0^{2\pi}\int_0^{2\pi} |e^{i\theta}-e^{i\theta'}|^{2(l-2)} (1-e^{i(\theta'-\theta)})$$

$$\ln\frac{|e^{i\theta}-e^{i\theta'}|^2}{2} f_1(e^{i\theta})\bar f_2(e^{i\theta'})\,d\theta\, d\theta'$$

(resp. $\langle\ldots,\ldots\rangle_q^- := -\langle\ldots,\ldots\rangle_q^+$) and H_+^q (H_-^q) as the completion of $C_+^l(S^1)$ $(C_-^{l-1}(S^1))$ with $\langle\ldots,\ldots\rangle_q^+$ $(\langle\ldots,\ldots\rangle_q^-)$. $\{\psi_k : k \geq 1\}$ $(\{\psi_k : k \leq -(l-1)\})$ is an orthogonal basis for H_+^q (H_-^q) . These representations belong to the discrete series, namely those with odd weights.

(4.5) Remark:

As we have seen in (3.6) (iv), the case $l=1$ $(q=0)$ for V^q is possible, too, and must lead to representations with highest or lowest weights $+1$ resp. -1 . Then the realisations of (4.4) no longer hold. But in (2.1) we have already constructed a (reducible) unitary representation V^0 . A short computation shows that $C_+^1(S^1)$ resp. $C_-^0(S^1)$ are indeed stable under the action of V^0. So we take their completions (now with respect to the canonical inner product of $L_2(S^1)$) for H_+^0 and H_-^0 . Because of their bivalent character these representations are sometimes called those of the *mock discrete series*.

We have now explicitly constructed realisations of irreducible unitary representations of $SU(1,1)$ for any possible (according to (3.6)) $q \in \mathbb{C}$, and so we have obtained a complete list (up to unitary equivalence) of all such representations of $SU(1,1)$.

There is another rather more usual realisation of the discrete series in the literature (see [27]):

Let $\mathcal{X}_n (n \geq 2)$ be the Hilbert space of all weakly measurable complex-valued functions f defined on the unit disc D for which the integral

$$\int_D f(z)\bar{f}(z)(1-|z|^2)^{n-2} d\mu_2 \qquad (\mu_2 \text{ Lebesgue measure})$$

exists. Then we define for any $g \in SU(1,1)$

$$(\Pi_g^n f)(z) = \frac{1}{(\alpha-\bar{\beta}z)^n} f\left(\frac{\bar{\alpha}z-\beta}{\alpha-\bar{\beta}z}\right) .$$

Then Π^n becomes a unitary representation with respect to the inner product given by

$$\langle f_1,f_2\rangle_n := \int_D f_1(z)\bar{f}_2(z)(1-|z|^2)^{n-2} d\mu_2 .$$

One easily shows that Π^n restricted to $\mathcal{K}_n := \{f \in \mathcal{H}_n : f$ holomorphic on $D\}$ is an irreducible representation of $SU(1,1)$ on $\mathcal{K}_n$.

Similarly Π^n, given by

$$(\Pi^n_g f)(z) = \frac{1}{(\bar{\alpha}-\beta\bar{z})^n} f\left(\frac{\bar{\alpha}z-\beta}{\alpha-\bar{\beta}z}\right) ,$$

defines an irreducible unitary representation of $SU(1,1)$ on $\tilde{\mathcal{K}}_n := \{f \in \mathcal{H}_n : f$ antiholomorphic on $D\}$.

It is not difficult to see that Π^n is of highest weight $-n$ and Π^n of lowest weight n. So according to our former classification they must be unitarily equivalent to some U^q resp. V^q. Indeed if we set $\phi_k := z^k (z \in D_1)$, which form obviously an orthogonal basis of $\mathcal{K}_n$, and if, for example, $n=2l$ with $l \in \mathbb{N}_+$, the mapping given by

$$L(\psi_k) := C \cdot \phi_{k-1} \qquad \text{(with a certain } C \in \mathbb{C})$$

can be extended to a unitary equivalence $L: \mathcal{K}_n \to H^q_+$ which commutes with Π^n and U^q $(q=2l-1)$. Similarly one can construct unitary equivalences for the other cases of the discrete series (for more details see [4]).

5. The First Order Cocycles of SU(1,1)

We are now able to construct all non-trivial first order cocycles of $SU(1,1)$ associated with unitary representations. As we have proved before all the irreducible unitary representations of $SU(1,1)$ can be induced from certain one-dimensional representations of the subgroup D.

The cocycles for such representations which are induced from unitary representations are completely classified by a theorem of Parthasarathy-Schmidt (see [20]):

(5.1) Theorem:

Let L be a unitary representation of a subgroup K of G, in a

Hilbert space V, let U^L be the induced representation of G (see section 2).

(i) If δ is a cocycle for U^L (due to the construction of the representation space of U^L for any $g \in G$ $\delta(g)$ is a square-integrable function from G/K to V), then there exists a Borel map $f:G/K \to V$ and a cocycle $\varepsilon : K \to V$, associated with the representation $k \mapsto \hat{L}_k = \tau(k)L_k$ of K on V, so that the following equation holds for any $g \in G$, a.e. $x(\mu)$:

$$\delta(g,x) = \{(U_g^L f)(x) - f(x)\} + \gamma(\rho(x))^{\frac{1}{2}} \varepsilon(\rho(x)^{-1} g \rho(g^{-1}x)) \qquad (*) \quad .$$

Here $\tau : K \to \mathbb{R}$ is the Mackey homomorphism (see [20], p.88) and γ is a uniquely determined function from G to $\mathbb{R}$ (see [20], p. 87).

(ii) Conversely, for any Borel map $f:G/K \to V$ and any cocycle ε of $\hat{L}$ such that the right hand side of (*) is a continuous map from G to $L_2(\mu,V)$, (*) yields a continuous cocycle for U^L.

(5.2) Remarks:

(i) If ε is a coboundary for $\hat{L}$, then $\delta(g,x)$ can be given only by the first part of (*) for some suitable Borel map f.

(ii) If f is not only Borel but also lies in $L_2(\mu,V)$, $(U_g^L f)(x) - f(x)$ is by definition a coboundary (and we are not interested in it); otherwise if f is Borel but not in $L_2(\mu,V)$ and $(U_g^L f)(x) - f(x)$ is in $L_2(\mu,V)$, the cocycle then is called a quasi-coboundary.

(iii) Out of the proof of (5.1), which is straightforward but somewhat length, we see that we do not necessarily need L to be unitary. But only in this case we are sure that the representation space of U^L only consists of functions.
If we induce from non-unitary representations as we have done in the sections 3 and 4, we only have a pre-Hilbert space of functions and we do not know what the completion looks like. So we can generalize 5.1 to the case that L is not unitary with the important restriction that (*) now no longer gives every

cocycle δ of U^L but only those which have only values in the pre-Hilbert space of functions. We shall see, however, that in our case of SU(1,1) this is no real restriction.

Instead of the "continuous" cocycle problem we are going to consider the "analytic" one. A first order cocycle δ is called an *analytic cocycle*, if δ is not only continuous but even analytic. It is evident from the cocycle equation that the range of an analytic cocycle is a subset of the analytic vectors of the associated representation. On the other hand Pinczon and Simon (see [23]) have shown that for a connected real Lie group the first cohomology group - continuous or analytic - is the same, this means that (up to a coboundary) every continuous cocycle is already an analytic one. So we only want to determine the analytic cocycles.

Therefore we must first describe the analytic vectors for our representations of SU(1,1) . Using a remark in Warner's book (see [30], p. 303) we see that for the representations of the principal and the complementary series the analytic vectors are at least continuous functions. For the discrete series Warner uses the realisation mentioned in (4.5) for the description of the analytic vectors. If we now take our unitary equivalence L (see (4.5)) it is not difficult to see that in our description the analytic vectors must be continuous functions, too.

So in any case the analytic vectors are in the pre-Hilbert spaces of functions we get by the induction (also from non-unitary representations). That is why the formula in (5.1) gives a complete solution of our cocycle problem.

Applying (5.1) to SU(1,1) we must first determine the function $\gamma: SU(1,1) \rightarrow \mathbb{R}$. A simple computation leads to

$$\gamma\left(\begin{bmatrix} \alpha & \beta \\ \beta & \alpha \end{bmatrix}\right) = |\alpha + \beta|^2 .$$

We now want to consider the second term of (*) . Therefore we need all the non-trivial cocycles associated with representations $\hat{L}_d = \tau(d) L_d$ of the subgroup D . The Mackey homomorphism turns out to be

$$\tau(d) = e^t ,$$

where t is the real parameter of the abelian part in the Iwasawa decomposition of d. Recalling that due to (3.6) we have induced all the unitary representations from one-dimensional representations of D given by the characters χ_1^q or χ_2^q we also get all the relevant L_d's for characters χ_1^q or χ_2^q with parameters which are just $q+1$ of the old q's. For this situation we get a complete classification of all first order cocycles by a lemma of Parthasarathy and Schmidt (see [20], p. 102):

(5.3) Lemma:

Let H be a topological group, let N be a normal and B an abelian subgroup of H such that $H=N\cdot B$ and $N\cap B=\{e\}$. Let χ be a one-dimensional non-trivial representation of H which is trivial on N.

Then $\varepsilon: H \to \mathbb{C}$ is a cocycle for χ iff

$$\varepsilon(nb) = \eta(n) + c(\chi(b)-1) ,$$

where c is a complex number and η is an additive continuous homomorphism from N to $\mathbb{C}$ with $\eta(b^{-1}nb)=\chi(b^{-1})\eta(n)$ for any $(n,b)\in N\times B$.

Obviously this lemma is applicable to our situation if we take N as the nilpotent subgroup of $SU(1,1)$ and $B=A\cdot Z$. Since we are only interested in non-trivial cocycles we can omit the second term in (5.3). As N is isomorphic to $(\mathbb{R},+)$ (see (1.2)), every continuous non-trivial homomorphism η from N to $\mathbb{C}$ is given by

$$\eta(n) = s\cdot\gamma ,$$

where s is the real parameter of n and γ is a complex number $\neq 0$. Let us first consider characters $\hat{\chi}_1^q(d)=(e^t)^{q+1}$ (identity on the centre). Then we get:

$$\eta(b^{-1}nb) = \chi(b^{-1})\eta(n) \qquad \forall\ (n,b)\in N\times B$$

$$\Rightarrow\ se^{-2t}\gamma = (e^{-t})^{q+1}s\gamma \qquad \forall\ (s,t)\in\mathbb{R}^2$$

$$\Rightarrow\ e^{-2t} = e^{-tq}e^{-t} \qquad \forall\ t\in\mathbb{R}$$

$$\Rightarrow\ q = 1 .$$

If we now take the characters χ_2^q (non-trivial on the centre) we immediately see that no non-trivial η can exist fulfilling the above condition. So the only case where a non-trivial cocycle ε for $\hat{L}$ exists is given for χ_1^1 . Inducing from this character we obtain (following section 3 and 4) two irreducible representations namely one with highest and one with lowest weight. Let us now consider U^1 as the reducible unitary representation on the direct sum of the two invariant subspaces H_+^1 and H_-^1 . Then returning to (5.1) we can explicitly compute

$$\hat{\delta}(g,x) = \gamma(\rho(x))^{\frac{1}{2}}\varepsilon(\rho(x)^{-1}g\rho(g^{-1}x)) \quad :$$

Because of the special form of $\rho(x)$ we get $\gamma(\rho x))^{\frac{1}{2}}\equiv 1$. For the second term in $\delta(g,x)$ we need the s-parameter of $\rho(x)^{-1}g\rho(g^{-1}x)$. From (1.3) we get after some computation:

$$s = \frac{1}{2i}(\alpha\bar{\beta}\frac{\bar{\alpha}x-\beta}{\alpha-\bar{\beta}x} - \bar{\alpha}\beta\frac{\alpha\bar{x}-\bar{\beta}}{\bar{\alpha}-\beta\bar{x}})$$

and thus

$$\delta(g,x) = \frac{\gamma}{2i}(\underbrace{\frac{\bar{\beta}x}{\alpha-\bar{\beta}x}}_{\delta_1(g)(x)} - \underbrace{\frac{\beta\bar{x}}{\bar{\alpha}-\beta\bar{x}}}_{\delta_2(g)(x)}) \ .$$

Obviously for any $g \in SU(1,1)$ $\delta_1(g)$ is continuous on S^1 . Besides this we have:

$$\delta_1(g)(x) = \frac{\bar{\beta}x}{\alpha-\bar{\beta}x} = \frac{\bar{\beta}}{\alpha}x\cdot\frac{1}{1-\frac{\bar{\beta}}{\alpha}x} = \sum_{\nu=1}^{\infty}(\frac{\bar{\beta}}{\alpha}x)^{\nu} \quad ,$$

for the geometric series $\sum_{\nu=0}^{\infty}(\frac{\bar{\beta}}{\alpha}x)^{\nu}$ converges to

$$\frac{1}{1-\frac{\bar{\beta}}{\alpha}x} \quad , \text{ as } |\frac{\bar{\beta}}{\alpha}x| = |\frac{\bar{\beta}}{\alpha}| < 1 \ .$$

So δ_1 maps $SU(1,1)$ into $C_+^1(S^1)$. Similarly we see that the range of δ_2 is contained in $C_-^1(S^1)$. That is why δ_1 is a cocycle for the irreducible representation U^1 (considered on H_+^1) and δ_2 one for U^1 (on H_-^1). These are the only cocycles which arise from non-trivial cocycles of the subgroup. Next we want to look at the quasi-coboundaries. As any cocycle for $SU(1,1)$ is obviously a cocycle for the maximal compact subgroup K , where every cocycle is a coboundary, we can without loss of generality postulate for a quasi-coboundary $\tilde{\delta}(g)(x)=(U_g^q f)(x)-f(x)$:

$$\tilde{\delta}(g) = 0 \qquad \forall\ g \in K$$

$$\Longleftrightarrow\ (U_g^L f)(x) = f(x) \qquad \forall\ g \in K \qquad \forall\ x \in S^1\ .$$

Let us first consider the quasi-coboundaries for the representations U^q . Then we have because of the special form of K :

$$(U_g^q f)(x)-f(x) = 0 \qquad \forall\ g \in K \quad \forall\ x \in S^1$$

$$\Longleftrightarrow\ f(e^{-2i\theta}x) = f(x) \qquad \forall\ \theta \in R \quad \forall\ x \in S^1\ .$$

For a fixed $x \in S^1$ $e^{-2i\theta}x$ can reach any value on S^1 , so f must necessarily be a constant function on S^1 . But for the principal and the complementary series the constant functions are in the representation spaces. So in these cases all the quasi-coboundaries are coboundaries and we are not interested in them.

But for no member of the discrete series constant functions belong to the representation space. So it might be possible that for a certain $q=2l-1$ $\tilde{\delta}(g)(x)=(U_g^q f)(x)-f(x)$ (with f constant on S^1) is a non-trivial cocycle for U^q . Therefore we have to investigate whether $\tilde{\delta}(g)$ is in the representation space of U^q for every $g \in SU(1,1)$. Using Cauchy's integral formula one can show that for every $l \geq 2$ there exists a $g \in SU(1,1)$ such that $\tilde{\delta}(g)$ (which must be a continuous function) has a Fourier expansion with a non-zero a_{l-1} . So $\tilde{\delta}$ can be a cocycle for U^q neither on H_+^q nor on H_-^q .

For $l=1$, however, these arguments do not hold. One can easily show that in this case

$$\tilde{\delta}(g)(x) = (U_g^1 f(x)-f(x) = C\cdot(\frac{1}{|\alpha-\bar{\beta}x|^2} - 1) \quad \text{(with a constant } C\text{)}$$

has no constant term in the Fourier expansion, so $\tilde{\delta}$ is a cocycle for U^1 on $H_+^1 \oplus H_-^1$. Decomposition into the irreducible parts, however, yields:

$$\tilde{\delta}(g)(x) = C\cdot(\frac{1}{|\alpha-\bar{\beta}x|^2} - 1) = C\cdot(\frac{\bar{\beta}x}{\alpha-\bar{\beta}x} + \frac{\beta\bar{x}}{\bar{\alpha}-\beta\bar{x}})\ .$$

So by the quasi-coboundary $\tilde{\delta}$ we get the same cocycles for the irreducible representations as before.

We will return to the consideration of the quasi-coboundaries for V^q. The postulation $\tilde{\delta}(g)=0$ on K then leads to:

$$e^{i\theta}f(e^{-2i\theta}x)-f(x) = 0 \qquad \forall\ \theta \in \mathbb{R} \quad \forall\ x \in S^1 .$$

Taking $\theta = \pi$ immediately shows that the last equation can only hold for $f \equiv 0$. So we have $\tilde{\delta} \equiv 0$.

Thus the only representations with (possible) non-trivial cohomology are given by U^1 on H^1_+ resp. H^1_-. It still remains to show that δ_1 and δ_2 are indeed non-trivial. But this is straightforward: An easy computation (see 5.5) yields $\|\delta_1(g)\|^2 = \|\delta_2(g)\|^2 = \gamma \log|\alpha|$ with some positive constant γ and $g = \begin{pmatrix} \alpha & \beta \\ \bar{\beta} & \bar{\alpha} \end{pmatrix} \in SU(1,1)$. So the mappings δ_1 and δ_2 are unbounded and this is according to a theorem of Parry and Schmidt (see [18]) necessary and sufficient for δ_1 and δ_2 to be non-trivial.

We do not yet know if $g \mapsto \delta_1(g)$ and $g \mapsto \delta_2(g)$ are analytic. But because of the equality of analytic and continuous cohomology we only need to show the continuity of the above mappings which is obtained by some simple ε-arguments. Summarizing the last section we have proved

(5.4) Theorem:

There are exactly two non-trivial first order cocycles associated with irreducible unitary representations, namely:

$$\delta_1 : SU(1,1) \to H^1_+ \qquad \text{given by}$$

$$\delta_1(g)(x) := \frac{\bar{\beta}x}{\alpha - \bar{\beta}x}$$

is the only non-trivial cocycle for U^1 on H^1_+ and

$$\delta_2 : SU(1,1) \to H^1_+ \qquad \text{given by}$$

$$\delta_2(g)(x) := \frac{\beta\bar{x}}{\bar{\alpha} - \beta\bar{x}}$$

is the one for U^1 on H^1_- .

For applying the results on positive definite functions (see Chapter V) we need the expressions $\mathrm{Im}\langle\delta(g_2),\delta(g_1^{-1})\rangle$ and $\langle\delta(g),\delta(g)\rangle$ for non-trivial cocycles. So we want to calculate $\langle\delta(g_2),\delta(g_1^{-1})\rangle$ first in the case $\delta=C\delta_1$ (with C a complex constant). We have to look at H^1_+ with its inner product $\langle\ldots,\ldots\rangle_1$ defined in (4.3). As δ_1 maps $SU(1,1)$ not only into H^1_+ but into $C^1_+(S^1)$ we only need to investigate $\langle h_1,h_2\rangle_1$ for elements of $C^1_+(S^1)$. So given h_1 and h_2 with their Fourier expansions (in the "classical" sense) $\sum_{k=1}^{\infty} a_k z^k$ resp. $\sum_{k=1}^{\infty} b_k z^k$ we get:

$$\langle h_1,h_2\rangle_1 = 8\pi^2 \sum_{k=1}^{\infty} \frac{1}{k} a_k \bar{b}_k \qquad (*)$$

(Note that the z^k form an orthogonal but not an orthonormal basis for H^1_+ with respect to $\langle\ldots,\ldots\rangle_1$!)
Now the computation of the Fourier coefficients turns out to be very simple, namely:

$$\delta(g)(z) = C\cdot\frac{\bar{\beta}z}{\alpha-\bar{\beta}z} = C\cdot \sum_{k=1}^{\infty} \left(\frac{\bar{\beta}}{\alpha}\right)^k z^k .$$

So we have from $(*)$:

$$\langle\delta(g_2),\delta(g_1^{-1})\rangle_1 = \pi^2|C|^2 \sum_{k=1}^{\infty} \frac{1}{k} \left(-\frac{\bar{\beta}_2\beta_1}{\alpha_2\alpha_1}\right)^k .$$

Using the formula $\sum_{k=1}^{\infty} \frac{1}{k} h^k = -\log(1-h)$ valid for any $h\in\mathbb{C}$, $|h|<1$ ($\log$ denotes the complex logarithm defined and differentiable in $\mathbb{C}\setminus]-\infty,0]$ with $\log 1 = 0$) we finally get:

$$\langle\delta(g_2),\delta(g_1^{-1})\rangle_1 = -8\Pi^2|C|^2 \log \frac{\alpha_1\alpha_2+\beta_1\bar{\beta}_2}{\alpha_1\alpha_2} .$$

Let us now turn to $\delta=C\cdot\delta_2$.

It is obvious from the definition of δ_2 and $\langle\ldots,\ldots\rangle_1$ that $\langle\delta_2(g_2),\delta_2(g_1^{-1})\rangle=\langle\delta_1(g_2),\delta_1(g_1^{-1})\rangle$. So we only have to take the complex conjugate of the last expression in that case.

It will be of special interest later on to look at the case $\delta=C\cdot(\delta_1+\delta_2)$

(δ is a cocycle for the reducible unitary representation U^1 on $H^1_+ \oplus H^1_-$). Because of the orthogonality of H^1_+ and H^1_- we get:

$$\begin{aligned}<\delta(g_2),\delta(g_1^{-1})>_1 &= |C|^2[<\delta_1(g_2),\delta_1(g_1^{-1})>_1+<\delta_2(g_2),\delta_2(g_2^{-1})>_1] \\ &= 2\ \mathrm{Re}\ \{|C|^2<\delta_1(g_2),\delta_1(g_1^{-1})>_1\}\ .\end{aligned}$$

Summarizing these results we get

(5.5) Lemma:

Let δ_1,δ_2 be the cocycles given in (5.4). For a given cocycle let

$$\begin{aligned} s(g_1,g_2) &:= \mathrm{Im}<\delta(g_2),\delta(g_1^{-1})> \\ \psi(g) &:= -\frac{1}{2}\,||\delta(g)||^2\ . \end{aligned}$$

Then we have for

(i) $\delta = C\delta_1$:
$$s(g_1,g_2) = -8\pi^2|C|^2\ \mathrm{Im}\ \log\frac{\alpha_1\alpha_2+\beta_1\bar{\beta}_2}{\alpha_1\alpha_2}$$
$$\psi(g) = -8\pi^2|C|^2\ \log|\alpha|$$

(ii) $\delta = C\delta_2$:
$$s(g_1,g_2) = 8\pi^2|C|^2\ \mathrm{Im}\ \log\frac{\alpha_1\alpha_2+\beta_1\bar{\beta}_2}{\alpha_1\alpha_2}$$
$$\psi(g) = -\ 8\pi^2|C|^2\ \log|\alpha|$$

(iii) $\delta = C(\delta_1+\delta_2)$:
$$s(g_1,g_2) \equiv 0$$
$$\psi(g) = -16\pi^2|C|^2\ \log|\alpha|$$

6. The First Order Cocycles of SL(2;$\mathbb{C}$)

In the last section of this chapter we want to give a brief summary of results concerning the first order cohomology of SL(2;$\mathbb{C}$) (for details see [7]). The arguments are widely analoguous to those we used for SL(2;$\mathbb{R}$), so we only give the results.

In [7] all the irreducible unitary representations of $SL(2;\mathbb{C})$ are determined:

(6.1). Theorem:

Let $m \in \mathbb{Z}$, $r \in \mathbb{C}$, $g = \begin{bmatrix} \alpha & \beta \\ \gamma & \delta \end{bmatrix} \in SL(2;\mathbb{C})$ $(\alpha\delta-\beta\gamma=1)$.

Define $(U_g^{m,r}f)(z)=|\alpha-\gamma z|^{m+ir-2}(\alpha-\gamma z)^{-m}f(\frac{\delta z-\beta}{\alpha-\gamma z})$ where f is a complex-valued square-integrable function (with respect to the standard Lebesgue measure on $\mathbb{C}$).

(i) For any $m \in \mathbb{Z}$, $r \in \mathbb{R}$ $U^{m,r}$ is an irreducible unitary representation of $SL(2;\mathbb{C})$ on the Hilbert space of square integrable functions with respect to the inner product

$$\langle f_1,f_2\rangle = \int_{\mathbb{C}} f_1(z)\bar{f}_2(z)\,dxdy \qquad \text{with} \quad z = x + iy .$$

(ii) For $m=0$, $r=i\rho$ with $0<\rho<2$ $U^{m,r}$ is an irreducible unitary representation of $SL(2;\mathbb{C})$ on the Hilbert space of all functions $f: \mathbb{C} \to \mathbb{C}$ satisfying

$$\int_{\mathbb{C}^2} |z_1-z_2|^{-2+\rho} |f(z_1)|\,|f(z_2)|\,dx_1dy_1dx_2dy_2 < \infty$$

with respect to the inner product

$$\langle f_1,f_2\rangle_\rho = \int_{\mathbb{C}^2} |z_1-z_2|^{-2+\rho} f_1(z_1)\bar{f}_2(z_2)\,dx_1dy_1dx_2dy_2 .$$

The representation under (i) constitute the principal series, those under (ii) the complementary series. Any irreducible unitary representation of $SL(2;\mathbb{C})$ is unitarily equivalent to one of the above forms.

For applying the theorem on cocycles of induced representations (see (5.1)) we have to show how all these representations may be induced. As the closed subgroup for our inducing cosntruction we take

$$T = \{ \begin{bmatrix} \lambda & 0 \\ \lambda z & \lambda^{-1} \end{bmatrix} : z \in \mathbb{C} , \lambda \in \mathbb{C}^* \}$$

noting that T is the semi-direct product of

$$N = \{ \begin{bmatrix} 1 & 0 \\ z & 1 \end{bmatrix} : z \in \mathbb{C}\} \quad \text{and} \quad A = \{ \begin{bmatrix} \lambda & 0 \\ 0 & \lambda^{-1} \end{bmatrix} : \lambda \in \mathbb{C}^* \} .$$

Then we obtain (see [7]):

(6.2) Lemma:

Let a character $L^{m,r}$ of T be given by

$$L^{m,r}\left(\begin{pmatrix} \lambda & 0 \\ \lambda z & \lambda^{-1} \end{pmatrix}\right) = e^{ir \log a} e^{imt}$$

with $\lambda = ae^{-it}$ $(t \in \mathbb{R}, a > 0)$. Then the representations $U^{m,r}$ given in (6.1) can be induced (as in section 2) from the corresponding $L^{m,r}$. The characters with $m \in \mathbb{Z}$, $r \in \mathbb{R}$ are the unitary ones.

Having shown that any quasi-coboundary of $SL(2;\mathbb{C})$ is a coboundary an analysis of the cocycle problem for the semi-direct product T finally gives the only non-trivial cocycle for $SL(2;\mathbb{C})$ (for details see [7]):

(6.3) Theorem:

There is precisely one non-trivial cocycle for $SL(2;\mathbb{C})$ namely

$$\delta(g,z) = C\cdot \frac{(\alpha\bar{z}+\gamma)(\bar{\alpha}-\bar{\gamma}\bar{z})+(\beta\bar{z}+\delta)(\bar{\beta}-\bar{\delta}\bar{z})}{(1+|z|^2)(|\alpha-\gamma z|^2+|\delta z-\beta|^2)}$$

with an arbitrary $C \in \mathbb{C}$.
This is associated with the representation $U^{m,r}$ with $m=2$, $r=0$ which belongs to the principal series.

For our applications on infinitely divisible projective representations we need the expression $\psi(g) = -\frac{1}{2}\|\delta(g)\|^2$ for the above cocycle.

Let $SU(2) = \left\{\begin{pmatrix} \alpha & \beta \\ -\bar{\beta} & \bar{\alpha} \end{pmatrix} : \alpha, \beta \in \mathbb{C}, |\alpha|^2+|\beta|^2 = 1\right\}$, then $SU(2)$ is a maximal compact subgroup of $SL(2;\mathbb{C})$.

We note that any $g \in SL(2;\mathbb{C})$ can be uniquely written as $g=k_1 \varepsilon k_2$ with $k_1,k_2 \in SU(2)$, $\varepsilon \in A$ with $\lambda \in \mathbb{R}$ and $\lambda > 1$. A straightforward but somewhat lengthy computation yields for a given $g \in SL(2;\mathbb{C})$:

$$\lambda = \frac{\sqrt{\mathrm{tr}\, g^* g + 2} + \sqrt{\mathrm{tr}\, g^* g - 2}}{2} \tag{1}$$

where tr denotes the trace of a matrix.

On the other hand we have (with $U = U^{2,0}$ for brevity):

$$\begin{aligned} \delta(g) = \delta(k_1 \varepsilon\, k_2) &= U_{k_1}\delta(\varepsilon k_2) + \delta(k_1) \\ &= U_{k_1}\delta(\varepsilon k_2) \qquad \text{(note that } \delta(k)=0 \text{ for any } k \in SU(2)\ !) \\ &= U_{k_1}[U_\varepsilon \delta(k_2) + \delta(\varepsilon)] \\ &= U_{k_1}\delta(\varepsilon)\ . \end{aligned}$$

Because of the unitarity of U we may then conclude:

$$\|\delta(g)\|^2 = \|U_{k_1}\delta(\varepsilon)\|^2 = \|\delta(\varepsilon)\|^2 .$$

But $\|\delta(\varepsilon)\|^2$ is computed in [7], namely:

$$\|\delta(\varepsilon)\|^2 = 2\pi(2t \coth t - 1) \tag{2}$$

where $t = \log\lambda$.
Combining (1) and (2) we finally get an expression for $\psi(g)$.

V. FURTHER RESULTS ON SEMI-SIMPLE LIE GROUPS

1. Kazdan's Result

We are now going to prove an important result due to Kazdan cf.[32]. This is a negative result in the sense that it asserts that a large class of groups only possesses a trivial cohomology. We are going to deal only with the case where the group G in question is a semi-simple Lie group. So in this chapter G will be a semi-simple Lie group unless otherwise specified.

In order to clarify the situation we first of all need to introduce the concept of a "conditionally s-positive function" (c.s.p. function). Although it is not clear at this stage why this should be necessary we ask the reader to have patience since the importance will become obvious in due course.

(1.1) Definition:

Let $\varphi : G \to \mathbb{C}$ and $s \in Z^2(G,\mathbb{R})$ be given such that

(i) φ is continuous, s is continuous

(ii) $\varphi(e) = 0$, $s(g,h) = -s(h^{-1},g^{-1})$, $s(e,g) = s(g,e) = 0$, $\forall (g,h) \in G\times G$

(iii) $\varphi(h^{-1}) = \overline{\varphi}(h) \quad \forall\, h \in G$

then the pair (φ,s) is called <u>conditionally s-positive</u> if

$$\sum_{k=1}^{n} \sum_{j=1}^{n} \alpha_k\overline{\alpha}_j \,[\varphi(g_j^{-1} g_k) + i\, s(g_j^{-1}, g_k)] \geq 0$$

$$\forall (\alpha_1,\dots,\alpha_n) \in \mathbb{C}^n \qquad \forall (g_1,\dots,g_n) \in G^n \qquad \text{with } \sum_{k=1}^{n} \alpha_k = 0$$

Remark:

If $s \equiv 0$ the above definition reduces to that of a normalized (i.e. 0 at the identity) conditionally positive function.

(1.2) Theorem cf. [6]:

For every c.s.p. pair (φ,s) there exists a pair (U,δ) of a representation U of G and a first order cocycle δ associated with it such that $\{\delta(g) : g\in G\}$ is total in the space in which U acts and

$$<\delta(g_1),\delta(g_2)> = \varphi(g_2^{-1}g_1) - \varphi(g_1) - \varphi(g_2^{-1}) + is(g_2^{-1},g_1) \qquad (*)$$

(U,δ) are here determined up to unitary equivalence. Conversely given (U,δ) as above we can find a pair (φ,s) which is c.s.p. and satisfies $(*)$. If (φ',s') is another pair satisfying $(*)$ then $\mathrm{Re}\varphi = \mathrm{Re}\varphi'$ and $s-s'$ is trivial.

We note that a pair (φ,s) satisfying $(*)$ and being c.s.p. may be obtained by setting

$$\varphi(g) := -\tfrac{1}{2}\|\delta(g)\|^2, \quad s(g_1,g_2) := \mathrm{Im}<\delta(g_2),\delta(g_1^{-1})>$$

We ask the reader to note the similarity to Theorem (2.2) in chapter II, where the above theorem was used as well.

The c.s.p. pairs are really a technical device to deal with conditionally positive (c.p.) functions. The connection between the two is given by

(1.3) Lemma:

There exists a bijection η between pairs (ψ,a) where ψ is c.p. and $a:G\to\mathbb{R}$ is continuous satisfying $a(g^{-1}) = -a(g)$ $\forall$ $g\in G$ and c.s.p. (φ,s) where s satisfies $s(g_1,g_2) = b(g_1g_2) - b(g_1) - b(g_2)$ for some continuous function $b : G\to\mathbb{R}$. This bijection is given by

$$\eta(\psi,a) := (\psi - ia,s) \quad \text{where}$$
$$s(g_1,g_2) = a(g_1g_2) - a(g_1) - a(g_2)$$

Proof:

Let (ψ,a) be as above, then we have

$$\sum_{k=1}^{n}\sum_{j=1}^{n} \alpha_k\bar{\alpha}_j\{\psi(g_j^{-1}g_k) - ia(g_j^{-1}g_k) + i[a(g_j^{-1}g_k) - a(g_k) - a(g_j^{-1})\}] =$$

$$\sum_{k=1}^{n}\sum_{j=1}^{n} \alpha_k\bar{\alpha}_j\{\psi(g_j^{-1}g_k) + i[a(g_j) - a(g_k)]\} =$$

$$\sum_{k=1}^{n}\sum_{j=1}^{n} \alpha_k\bar{\alpha}_j\ \psi(g_j^{-1}g_k) \geq 0$$

if $\sum_{k=1}^{n} \alpha_k = 0$ since ψ is c.p.

Hence $(\psi - ia,s)$, as above, is c.s.p. and thus η is well defined.

<u>η is surjective:</u> Let a c.s.p. pair (φ,s) with $s(g_1,g_2) = b(g_1g_2) - b(g_1) - b(g_2)$ be given then it is easy to see that

$$\eta(\varphi + ib, b) = (\varphi, s)$$

with $\varphi + ib$ c.p.

<u>η is injective:</u>

Suppose that two pairs (φ_1, s_1), (φ_2, s_2) are given. Then
if $s_1(g_1, g_2) = a_1(g_1 g_2) - a_1(g_1) - a_1(g_2)$ and
$s_2(g_1, g_2) = a_2(g_1 g_2) - a_2(g_1) - a_2(g_2)$
we have $s_1 \equiv s_2$ iff
$a_1(g) = a_2(g) + \beta(g)$
where $\beta : G \to \mathbb{R}$ is a continuous homomorphism. But G, being semi-simple is its own commutator subgroup and we must have $\beta \equiv 0$. Thus $a_1 \equiv a_2$. then if $\varphi_1 \equiv \varphi_2$ we must also have $\varphi_1 + ia_1 \equiv \varphi_2 + ia_2$ q.e.d

Consider now quadruples (H, U, δ, a) where U is a unitary representation of G in H, δ is a first order cocycle associated with U such that $\{\delta(g) : g \in G\}$ is total in H, $a : G \to \mathbb{R}$ is continuous and satisfies

$$a(g_1 g_2) - a(g_1) - a(g_2) = \mathrm{Im}\langle\delta(g_2), \delta(g_1^{-1})\rangle$$

We call two such quadruples (H, U, δ, a) and (H', U', δ', a') isomorphic, cf. [32], if there exists a unitary operator $A : H \to H'$ satisfying

(i) $U_g = AU'_g A^{-1}$ $\quad \forall\, g \in G$

(ii) $A\delta(g) = \delta'(g)$

and if $a \equiv a'$.

In view of (1.2) and (1.3) we can now state

(1.4) Theorem:

There exists a bijection between isomorphism classes (H, U, δ, a) as above and c.p. functions Ψ. This is obtained by setting

$$\Psi(g) := -\tfrac{1}{2}\|\delta(g)\|^2 + ia(g)$$

Moreover Ψ then satisfies

$$\Psi(g_2^{-1} g_1) - \Psi(g_1) - \Psi(g_2^{-1}) = \langle\delta(g_1), \delta(g_2)\rangle \quad \forall\; g_1, g_2 \in G\times G$$

It is now possible to formulate a necessary and sufficient condition for a first order cocycle to be a coboundary in terms of c.p. functions. In fact we have

(1.5) Lemma:

The cocycle δ associated with the c.p. function via (1.4) is a coboundary iff

$$\Psi(g) = f(g) - f(e)$$

where f is a positive definite function (nit necessarily normalized at e).

Proof:

Let $\Psi(g) = f(g) - f(e)$ where f is positive definite and let (H,U,δ,a) be the quadruple associated with Ψ via (1.4). Then we have

$$||\delta(g)||^2 = -2\mathrm{Re}(f(g) - f(e))$$

But f, being positive definite, is bounded and thus δ is bounded. Hence δ must be a coboundary.

If on the other hand $\delta(g) = U_g x - x$ (i.e. δ is a coboundary), then the equation

$$\begin{aligned} a(g_1g_2) - a(g_1) - a(g_2) &= \mathrm{Im}\langle U_{g_2}x - x, U_{g_1^{-1}}x - x\rangle \\ &= \mathrm{Im}\langle U_{g_1g_2}x,x\rangle - \langle U_{g_1}x,x\rangle - \langle U_{g_2}x,x\rangle \end{aligned}$$

is obviously satisfied, if we set

$$a(g) := \mathrm{Im}\langle U_g x,x\rangle .$$

This solution is, a priori, only determined up to an additive homomorphism but since G was assumed semi-simple it is unique.

Consequently we obtain:

$$\begin{aligned} \Psi(g) &= -\tfrac{1}{2}||\delta(g)||^2 + ia(g) = \\ &\langle U_g x,x\rangle - \langle x,x\rangle \end{aligned}$$

Hence $\Psi(g) = \varphi(g) - \varphi(e)$ with

$\varphi(g) = \langle U_g x,x\rangle$ q.e.d.

It will turn out that (1.5) is of crucial importance in the proof of Kazdan's Result.

We need two further results:

(1.6) Theorem (cf. [32]):

$\Psi: G \to \mathbb{C}$ is c.p. iff $\exp t\Psi$ is positive definite $\forall t \in \mathbb{R}$ with $t > 0$.

This result is fairly easy to prove but what we really need is a more or less trivial corollary namely

(1.7) Corollary:

If Ψ is c.p. then there exists a sequence $\{\Psi_n\}$ with $\Psi_n(g) = f_n(g) - f_n(e)$ and f_n positive definite $\forall n \in \mathbb{N}$ such that $\Psi(g) = \lim \Psi_n(g)$ where the convergence is uniform over compact subsets of G .

Proof:

We may for example take

$$f_n(g) = ne^{\frac{1}{n}\Psi(g)} \quad .$$

(This is arrived at of course by considering

$$\frac{d}{dt}\left(e^{t\Psi}\right)\Big|_{t=0} \quad .)$$

q.e.d.

We are now ready to state and prove the very powerful result due to Kazdan.

(1.8) Theorem:

Let G be a separable semi-simple Lie group. Suppose that G possesses the property (T) of Kazdan, i.e. suppose that the 1-dimensional trivial representation is isolated in $\tilde{G}$, the space of equivalence classes of unitary representations, furnished with the Fell topology.

Then the first order cohomology group of G is trivial.

Proof:

According to (1.5) it is sufficient to prove that every c.p. function ψ may be written as $\psi(g) = f(g) - f(e)$ where f is positive definite.

Because of (1.7) we know that

$$\psi(g) = \lim_n [f_n(g) - f_n(e)] = \lim_n \psi_n(g)$$

the limit being uniform over compact subsets and f_n being positive definite $\forall n \in \mathbb{N}$.

Suppose that $f_n(g) = \langle U_g^{(n)} x^{(n)}, x^{(n)} \rangle$ where $g \mapsto U_g^{(n)}$ is a unitary representation of G in a Hilbert space $H^{(n)}$ and $x^{(n)}$ is a cyclic vector in $H^{(n)}$.

It is clear that we may without loss of generality assume that no $U^{(n)}$ contains the trivial representation: If $U^{(n)}$ contained the trivial representation we could consider the orthogonal projection onto the invariant subspace, it thus being obvious that from the trivial representation no contribution to $\psi_n(g)$ could be made.

Now suppose that the sequence $f_n(e)$ is not bounded. Then we can find a subsequence f_{n_i} such that $\lim_{i\to\infty} f_{n_i}(e) = +\infty$.

Hence $\lim_{i\to\infty} \dfrac{f_{n_i}(g) - f_{n_i}(e)}{f_{n_i}(e)} = 0$ uniformly over compact sets.

That is to say that $\dfrac{f_{n_i}(g)}{f_{n_i}(e)}$ converges uniformly to 1 over compact sets and this implies that $U^{(n_i)}$ converges to the trivial 1-dimensional representation in $\tilde{G}$. But this is not possible, since G was assumed to possess property (T), unless $U^{(n_i)}$ contains the trivial representation from a certain stage.

Thus we have a contradiciton and $\{f_n(e)\}$ is bounded. So we may find a convergent subsequence $\{f_{n_j}(e)\}$ and set

$$\lim_{j\to\infty} f_{n_j}(e) = 1 .$$

Then $\lim_{j\to\infty} f_{n_j}(g) = \Psi(g) + 1$ and hence

$f(g) = \Psi(g) + 1 \Rightarrow$

$\Psi(g) = f(g) - f(e)$ q.e.d.

In order to see how far - reaching this result really is we follow [32]. We need to study spherical functions.

2. Spherical Functions

It will turn out that it is easier to check property (T) not on the space of representations directly but rather consider their associated "expectation values". These will turn out to be just the spherical functions.

First of all, however, let us proceed by proving a couple of technical lemmas which will make life easier in the future. We note that in (1.4) $\Psi(g)$ is just given by $\Psi(g) = -\frac{1}{2}\|\delta(g)\|^2$ if the cocycle is such that $\mathrm{Im}\langle\delta(g_2), \delta(g_1^{-1})\rangle = a(g_1g_2) - a(g_1) - a(g_2)$ is identically zero which implies that a is identically zero. We deduce easily

(2.1) Lemma:

There exists a bijection between isomorphism classes of triples (H,U,δ) where U is a unitary representation of G in H and δ a first order cocycle associated with U, satisfying

$$\mathrm{Im}\langle\delta(g_2), \delta(g_1^{-1})\rangle \equiv 0$$

and real-valued c.p. functions Ψ given by

$$(H,U,\delta) \mapsto \Psi(g) := -\tfrac{1}{2}\|\delta(g)\|^2$$

We denote the triple associated with Ψ by $(H_\Psi, U_\Psi, \delta_\Psi)$.

The point of (2.1) is that one can create somewhat artificially precisely that situation:

Suppose that (H, U, δ) , as above, is given.

Define a new Hilbert space $\overline{H}$ by $\overline{H} = H$ as an additive group.

Multiplication by a scalar is then given by

$$(\lambda, x) \mapsto \overline{\lambda} x \quad \forall \lambda \in \mathbb{C} \ \forall x \in H$$

and the scalar product in $\overline{H}$ is defined by

$$\langle x_1, x_2 \rangle_{\overline{H}} := \langle x_2, x_1 \rangle_H$$

We can then define a representation $\overline{U}$ and a first order cocycle $\overline{\delta}$ associated with it in $\overline{H}$ by setting:

$$\overline{U}_g x := U_g x \qquad \forall\ x \in H \quad \forall\ g \in G$$
$$\overline{\delta}(g) := \delta(g) \qquad \forall\ g \in G$$

Then we obtain easily:

(2.2) Lemma:

Taking

$$\delta_\psi(g) = \delta(g) \oplus \overline{\delta}(g) \qquad H \oplus \overline{H}$$

H_ψ as the $U \oplus \overline{U}$-stable subspace of $H \oplus \overline{H}$ generated by $\delta_\psi(G)$ and U_ψ as the subrepresentation of $U \oplus \overline{U}$ in H_ψ we obtain a triple $(H_\psi, U_\psi, \delta_\psi)$ associated with a real-valued c.p. function ψ given by

$$\psi(g) = -\ ||\delta(g)||^2 = -\ \tfrac{1}{2}||\delta_\psi(g)||^2$$

The crucial fact in (2.2) is that if δ_ψ is a coboundary in H_ψ then by taking the projection onto H we see that δ must be a coboundary also.

We are now ready to investigate the spherical functions. We consider G , a semi-simple connected Lie group with finite centre (unimodular). We denote by K a maximal compact subgroup. The commutative Banach algebra (with convolution) of integrable biinvariant functions with respect to K will be denoted by $L^1(K\backslash G/K)$.

(2.3) Definition:

A spherical function on G relative to K is a nonzero solution ω of the functional equation

$$\int_K \omega(g_1 k g_2)\, dk = \omega(g_1)\,\omega(g_2) \quad \forall\, g_1, g_2 \in G$$

where dk denotes the normalized Haar measure on K. A zonal spherical function ω (z.s. function) is biinvariant under K and satisfies $\omega(e) = 1$. We denote the set of positive definite z.s. functions by Ω.

We may now regard Ω as a part of the spectrum of the commutative Banach algebra $L^1(K\backslash G/K)$. Then the UCC topology and the topology of the spectrum coincide for Ω.

In fact it is shown in [33] that for the set of positive defintite functions which are 1 at the identity the UCC topology and the weak- $*$ topology for $L^\infty(G) = (L^1(G))'$ are the same. We note that if $\omega \in \Omega$ and $f \in L^1(G)$ we have

$$\int_G \omega(g) f(g)\, dg = \int_G \omega(g)\, {}^K f^K(g)\, dg$$

where ${}^K f^K \in L^1(K\backslash G/K)$ and is obtained from f by setting

$${}^K f^K(g) := \int_{K\times K} f(kgk')\, dk\, dk' .$$

Thus the weak- $*$ topology and the topology of the spectrum of $L^1(K\backslash G/K)$ coincide. Hence we obtain the stated result.

We now need to exhibit the connection between certain representations of G and the elements of Ω. The following theorem is well known (note the analogy to the GNS-construction):

(2.4) Theorem:

To each element $\omega \in \Omega$ we can associate an irreducible unitary representation U_ω in a Hilbert space H_ω and a vector x_ω such that

$$< U_\omega(g)x_\omega, x_\omega > = \omega(g) \quad , \quad U_\omega(k)x_\omega = x_\omega \quad \forall\ k \in K$$

Moreover, up to unitary equivalence, U_ω and x_ω are unique.

Thus we obtain a bijection Θ between Ω and equivalence classes of irreducible unitary representations having a nontrivial K-invariant vector (representations of class I).

We need

(2.5) Lemma [32]:

The mapping $\Theta : \Omega \to \hat{G}$ is a homeomorphism of Ω onto its image. This is open in $\hat{G}$.

Proof:

We show first that the image is open. We denote by $\tilde{K}$ the equivalence classes of unitary representations of K and again furnish $\tilde{K}$ with the Fell topology. Then, by restriction, we obtain a continuous map

$$R : \hat{G} \to \tilde{K} .$$

We consider

$F := \{U \in \tilde{K} : U$ doesn't contain the trivial representation of $K\}$.

K, being compact, has the property (T) of Kazdan. Hence the set F is closed. Thus $R^{-1}(F)$ is closed also and $\Theta(\Omega)$, being the complement of $R^{-1}(F)$ in $\hat{G}$ must be open.

Θ will obviously be continuous (for the topology of $\hat{G}$ see e.g. [33]). We show that Θ^{-1} is continuous.

Let $\hat{U}$ denote the equivalence class of U in $\hat{G}$. Suppose that $\lim_{i \in I} \hat{U}_i = \hat{U}$ where $\hat{U}_i$ is a net converging to $\hat{U}$ in $\hat{G}$.

Then there exists a net $\{\varphi_i\}_{i \in I}$ of positive definite functions and a net $\{x_i\}_{i \in I}$ of vectors in $H(U_i)$ where the representations U_i act such that

$$\varphi_i(g) = < U_i(g)x_i, x_i > .$$

Furthermore if $\omega = \Theta^{-1}(\hat{U})$ we have

$$\lim_{i \in I} \varphi_i(g) = \omega(g) \qquad \text{uniformly over compact subsets.}$$

If ${}^K\varphi_i{}^K(g) := \int_{K \times K} \varphi_i(kgk')\, dk\, dk'$, we have

$${}^K\varphi_i{}^K(g) = \langle U_i(g) P_i x_i, P_i x_i \rangle$$

where P_i is the projection onto the K invariant vectors in $H(U_i)$, i.e.

$$P_i x = \int_K U_i(k) x\, dx$$

Moreover ${}^K\varphi_i{}^K$ converges uniformly over compact sets to ${}^K\omega^K = \omega$.

There exists an i_o such that for $i \geq i_o$ we must have ${}^K\varphi_i{}^K(e) > 0$ and thus we can then normalize the ${}^K\varphi_i{}^K$ by taking

$\dfrac{{}^K\varphi_i{}^K}{{}^K\varphi_i{}^K(e)} = \omega_i$ and this then is obviously equal to $\Theta^{-1}(\hat{U}_i)$. Also we have $\lim_{i \in I} \omega_i = \omega$. Hence Θ^{-1} is continuous. q.e.d.

Following [32] further we denote by ε_G the trivial representation (1-dimensional!) and by $\mathfrak{1}$ the corresponding spherical function. Then we have

(2.6) Lemma:

If U is an irreducible unitary representation of G which is non-trivial and of class I then every first order cocycle associated with U is a coboundary. Here G is supposed to be a semi-simple connected Lie group with finite centre.

Proof:

Let δ be a first order cocycle for U . Then δ restricted to K must be a coboundary. Hence δ modulo a coboundary must be zero on K . So without loss of generality we are going to assume that δ is zero on K . Let $C_o(K\backslash G/K)$ denote the continuous biinvariant functions on

G with compact support. This is a commutative algebra (furnished with convolution). We suppose that U is acting in the Hilbert space H and define a map

$$B : C_o(K\backslash G/K) \longrightarrow H \quad \text{by}$$

$$B : \quad f \longmapsto \int_G f(g)\,\delta(g)\,dg$$

Let H^K denote the subspace of K-invariant vectors in H , with

$P := \int_K U(k)\,dk$ the projection onto H^K .

Then for all $f,g \in C_o(K\backslash G/K)$ we have

$$PB(f*h) = (\int_G f(g)\,dg)PB(h) + (\int_G h(g)U(g)\,dg)PB(f)$$

as is easily seen using the cocycle identity and invariance of Haar measure.

Since, by assumption, U is nontrivial there exists $f_o \in C_o(K\backslash G/K)$ with

$$\int_G f_o(g)\,dg \neq \int_G f_o(g)U(g)\,dg\ .$$

Utilizing the fact that $C_o(K\backslash G/K)$ is commutative we obtain:

$$PB(h) = [\int_G h(g)\,dg - \chi(h)]\ \frac{PB(f_o)}{\int_G f_o(g)\,dg - \chi(f_o)}$$

where we have set

$$\chi(h) = \int_G h\,(g)U\,(g)\,dg$$

Notice here that H^K is invariant under $\chi(h)$ and hence (H^K being 1-dimensional) $\chi(h)$ must be a scalar on H^K .

Now set

$$x = \frac{PB(f_o)}{\int_G f_o(g)\, dg - \chi(f_o)}$$

Then

$$PB(h) = [\int_G h(g)\, dg - \chi(h)]x$$

So defining $\quad \delta'(g) := \delta(g) + U(g)x - x$

and $\quad B'(f) := \int_G f(g)\, \delta'(g)\, dg$

we see that δ' is zero on K and that

$$PB' = 0$$

We note that

$$P \int_G h(g)\, \delta'(g)\, dg = PB'({}^K h^K)$$

for every continuous function h with compact support and hence that

$$P \int_G h(g)\, \delta'(g)\, dg = 0$$

for every continuous function h with compact support. Thus $P\delta' = 0$. Hence $\{\delta'(g) : g \in G\}$ can't be total in H. However, the subspace generated by $\{\delta'(g) : g \in G\}$ is invariant under U which was assumed to be irreducible. Hence it must be zero and $\delta = -(U_g x - x)$ is a coboundary. q.e.d.

We need some further information on c.p. functions:

(2.7) Lemma [32]:

Let (H, U, δ, a) correspond to the c.p. function ψ as in (1.4). Let δ be zero on K. Then ψ satisfies

$$\int_K \psi(g_1 k g_2)\, dk = \psi(g_1) + \psi(g_2)$$

iff $U|_K$ does not contain the trivial representation of K.

Proof:

$$\int_K < U(k)\,\delta(g_1), \delta(g_2) > dk = \int_K < \delta(kg_1), \delta(g_2) > dk$$

$$= \int_K [\Psi(g_2^{-1}kg_1) - \Psi(kg_1) - \Psi(g_2^{-1})]\ dk$$

$$= \int_K [\Psi(g_2^{-1}kg_1) - \Psi(g_1) - \Psi(g_2^{-1})]\ dk$$

Note here that $\delta \equiv 0$ on K implies that a must be a continuous homomorphism and hence $a \equiv 0$ on K since K is compact. Then Ψ is biinvariant under K in this case.

$U|_K$ doesn't contain the identity iff

$$\int_K U(k)\ dk = 0 \quad \text{iff} \quad \int_K < U(k)x, x' > dk = 0 \quad \forall\, x, x' \in H$$

$$\Leftrightarrow \int_K < U(k)\,\delta(g_1), \delta(g_2) > dk = 0 \qquad \forall\, g_1, g_2 \in G$$

since $\{\delta(g) : g \in G\}$ is total in H. q.e.d.

We also need the following theorem due to Gangolli [35] on real valued c.p. functions which are biinvariant under K:

(2.8) Theorem:

Let Ψ be a real-valued c.p. function biinvariant under K. Then we can write Ψ uniquely as

$$\Psi(g) = Q(g) + \int_{\Omega - \{\mathfrak{1}\}} (\omega(g) - 1)\ d\mu(\omega)$$

where $Q : G \to \mathbb{R}$ is a function and μ a measure on $\Omega - \{\mathfrak{1}\}$ such that

a) μ is a σ-finite positive measure which is invariant under the involution $\omega \mapsto \bar{\omega}$. The μ measure of the complement of an open neighbourhood of $\mathfrak{1}$ is finite.

b) If V is a compact neighbourhood of e in G with $K V K \quad V$ and $Q_V(\omega)$ defined on Ω by

$$Q_V(\omega) := \int_V (1 - \mathrm{Re}\, \omega(g))\, dg \Big/ \int_V dg$$

we have
$$\int_\Omega Q_V(\omega)\, d\mu(\omega) < \infty$$

c) Q is a continuous function of the form

$$Q(g) = \lim_{r \to \infty} \int_{V_r} (\omega(g) - 1)\, d\nu_r(\omega) \qquad g \in G$$

where $\{V_r\}$ is a decreasing sequence of compact neighbourhoods of $\mathbb{1}$ satisfying $\bigcap_r V_r = \{\mathbb{1}\}$ (Since Ω is locally compact the V_r are a neighbourhood basis for $\mathbb{1}$) and ν_r is a sequence of bounded positive measures which are invariant under $\omega \mapsto \bar{\omega}$

If μ satisfies a) and b) and Q c) we have:

Q and $\displaystyle\int_{\Omega - \{\mathbb{1}\}} (\omega(g) - 1)\, d\mu(\omega)$ are

real-valued c.p. functions and

$$\int_K Q(g_1 k g_2^{-1})\, dk + \int_K Q(g_1 k g_2)\, dk =$$

$$2[\, Q(g_1) + Q(g_2)]$$

We set $\displaystyle \Psi(g) := \int_{\Omega - \{\mathbb{1}\}} (\omega(g) - 1)\, d\mu(\omega)$

and realize the triple $(H_{\Psi_1}, U_{\Psi_1}, \delta_{\Psi_1})$ whose existence is guaranteed by (2.1). ω, being positive defintite, we consider the triple $(F_\omega, U_\omega, x_\omega)$ where

$$\omega(g) = \langle U_\omega(g) x_\omega, x_\omega \rangle$$

and U_ω acts in H_ω .

We construct the direct integral $\int_{\Omega-\{\mathbb{1}\}}^{\oplus} H_\omega \, d\mu(\omega)$ and set

$$\delta_{\Psi_1}(g) := \int_{\Omega-\{\mathbb{1}\}}^{\oplus} [U_\omega(g)x_\omega - x_\omega]d\mu(\omega)$$

Then δ is well defined since $\omega(g) - 1$ is integrable with respect to μ. Moreover δ is a first order coboundary or a limit of coboundaries for the representation $\int_{\Omega-\{\mathbb{1}\}}^{\oplus} U_\omega d\mu(\omega)$ and it's continuous since Ψ_1 is.

If we denote by H_{Ψ_1} the subspace of $\int_{\Omega-\{\mathbb{1}\}}^{\oplus} H_\omega d\mu(\omega)$ generated by $\{\delta_{\Psi_1}(g) : g \in G\}$ and by U_{Ψ_1} the subrepresentation of $\int_{\Omega-\{\mathbb{1}\}}^{\oplus} U_\omega d\mu(\omega)$ in H_{Ψ_1} then we have the required triple since

$$\Psi_1(g) = -\frac{1}{2}||\delta_1(g)||^2 = \int_{\Omega-\{\mathbb{1}\}}[\omega(g) - 1] \, d\mu(\omega)$$

Thus we see that the function Ψ_1 in the above theorem arises from a coboundary or a limit of coboundaries. In order to gain more information on Q we first need:

(2.9) Lemma [32] :

Let Ψ_o be real-valued and c.p. suppose that $\Psi_o = \Psi_1 + \Psi_2$ where Ψ_1 and Ψ_2 are real-valued, c.p., not identically zero. Let $(H_{\Psi_i}, U_{\Psi_i}, \delta_{\Psi_i})$ be the triples associated to Ψ_i $(i = 0,1,2)$ via (2.1).
Then there exists a nonzero intertwining operator $T_i : H_{\Psi_o} \to H_{\Psi_i}$ $i = 1,2$ between U_{Ψ_o} and U_{Ψ_i} sending δ_{Ψ_o} to δ_{Ψ_i} ; since $\{\delta_{\Psi_i}(g) : g \in G\}$ is total in H_{Ψ_i} there exists a subrepresentation of U_{Ψ_o} which is equivalent to U_{Ψ_j} .

Proof:

We consider $H_{\Psi_1} \oplus H_{\Psi_2}$ and $\delta: G \to H_{\Psi_1} \oplus H_{\Psi_2}$ defined by

$$\delta(g) := \delta_{\Psi_1}(g) \oplus \delta_{\Psi_2}(g)$$

If H then denotes the subspace of $H_{\Psi_1} \oplus H_{\Psi_2}$ which is generated by $\{\delta(g) : g \in G\}$ and U the subrepresentation of $U_{\Psi_1} \oplus U_{\Psi_2}$ in H, then (H,U,δ) is a triple associated with Ψ_o and the statement of the lemma is an immediate consequence. q.e.d.

We are now in a position to describe the function Q appearing in (2.8) more accurately. The result is indeed what we would expect:

(2.10) Lemma [32] :

Let U be an irreducible, nontrivial representation of G. If there exists a nontrivial first order cocycle δ associated with U (which we can assume to be zero on K) the function $Q := \|\delta\|^2$ satisfies condition c) in (2.8) and there exists a subrepresentation of $U \oplus \overline{U}$ equivalent to U_Q, where (H_Q, U_Q, δ_Q) is a triple associated with Q.

Proof:

(2.6) says that U can't be of class I. Let δ be a nontrivial cocycle for U (supposed to be zero on K) and set $\Psi(g) = -\|\delta(g)\|^2$. We write Ψ as $\Psi = \Psi_1 + Q$ as in (2.8).

According to (2.9) and our discussion after (2.8) we may deduce that if $\Psi_1 \not\equiv 0$ a subrepresentation of U_Ψ is unitarily equivalent to a subrepresentation of $\int^{\oplus}_{\Omega - \{\mathbf{1}\}} U_\omega \, d\mu(\omega)$. But we also have, according to our technical result (2.2) that U_Ψ is equivalent to a subrepresentation of $U \oplus \overline{U}$. But the subrepresentations of $U \oplus \overline{U}$ for U irreducible are just $U, \overline{U}$, and $U \oplus \overline{U}$.

As we have noted above a cocycle for $\int^{\oplus}_{\Omega - \{\mathbf{1}\}} U_\omega \, d\mu(\omega)$ is either a coboundary or a limit of coboundaries.

Hence the same must be true for U_Ψ which is equivalent to a subrepresentation of U_{Ψ_1}. However, $U \simeq U \oplus \overline{U}$, U or $\overline{U}$ implies immediately that any cocycle of U must be a coboundary or a limit of coboundaries. If it is not a coboundary then U must contain the trivial representation. Then $U|_K$ must contain the trivial representation and hence U must be of class I. Contradiction! Hence $\Psi = Q$ and a subrepresentation of $U \oplus \overline{U}$ is unitarily equivalent to U_Q according to our technical result (2.2). q.e.d.

We need a further lemma which will be extremely useful in the sequel.

(2.11) Lemma [32]:

Let G be a semi-simple, real, connected Lie group with centre $Z(G)$. Let U be an irreducible representation of G in H admitting a nontrivial cocycle. Then U is identity on $Z(G)$. Moreover let U' be the representation of $G'=G/Z(G)$ defined by U, then the cohomology of G with respect to U (denoted by $H^1(G,U)$) is isomorphic to $H^1(G',U')$.

Proof:

If U restricted to $Z(G)$ isn't trivial then $H'(Z(G),U) = O$. But then, since

$$\begin{aligned} U(z)\delta(g) &= \delta(zg) \\ &= \delta(gz) & \forall\ g,z \in GxZ(G) \\ &= U(g)\delta(z) + \delta(g) = \delta(g) & \forall\ \delta \in Z'(G,U) \end{aligned}$$

and since U restricted to $Z(G)$ doesn't contain the trivial representation, we see that $H^1(G,U) = O$ also. But this is the required contradiction and so U must be trivial on $Z(G)$.

We now have an exact sequence

$$O \to H^1(G',U') \xrightarrow{W} H^1(G,U) \xrightarrow{W'} \mathrm{Hom}_G(Z(G),H)$$

where $\mathrm{Hom}_G(Z(G),H)$ stands for the set of all continuous homomorphisms $f:G \to H$ satisfying

$$f(gzg^{-1}) = f(z) = U(g)f(z) \qquad \forall (g,z) \in GxZ(G)$$

Let $p: G \to G/Z(G)$ be the canonical projection, then W and W' are obtained by defining first

$$\begin{aligned} V: Z^1(G',U') &\to Z^1(G,U) \\ \delta &\mapsto \delta \circ p \end{aligned}$$

and $\quad V': Z^1(G,U) \to \mathrm{Hom}_G(Z(G),H)$

$$\delta \mapsto \delta|_{Z(G)}$$

W and W' are then obtained by passing to the quotient. If

$Hom_G(Z,H) \neq 0$, U must be trivial and thus

$$H^1(G',U') = H^1(G,U) = 0$$

because G is semi-simple. Thus $Hom_G(Z(G),H) = 0$ and so W is an isomorphism. q.e.d.

Remark:

Although fairly trivial to prove (2.11) is pretty useful, as will be seen in the next theorem already: It allows us to restrict our considerations to groups with finite centre. Moreover it is used in the explicit proof of (3.3) below given in [32] .

However, let us now use the result due to Kazdan in order to show that a large class of semi-simple Lie groups possesses only a trivial cohomology.

(2.12) Theorem [32] :

Let G be a semi-simple, real, connected Lie group with Lie algebra different from so(n;1) and su(n;1) then its first cohomology group is trivial.

Proof:

We may restrict ourselves to the case where G has finite centre (2.11). And it then suffices to show that G possesses Kazdan's property (T). If G has rank ≥ 2 this result is known [37] . Suppose now that G has rank = 1 and Lie algebra different from so(n;1), su(n;1) . Let KAN be an Iwasawa decomposition of G and let $\mathfrak{a}$ be the Lie algebra of A . Again let Ω be as in (2.3). According to [12] p. 428 the elements of Ω are parametrized by $\mathfrak{a}^*_{\mathbb{C}}$ (the complexification of the dual of $\mathfrak{a}$). So we may identify Ω with a part of $\mathfrak{a}^*_{\mathbb{C}}$. The topology corresponding to the natural topology of $\mathfrak{a}^*_{\mathbb{C}}$ is called the topology of parameters. That topology coincides (cf. [36]) with the topology induced in Ω by the natural topology in the spectrum of the commutative Banach algebra $L^1(K\backslash G/K)$. (Ω is here being regarded as part of the spectrum of $L^1(K\backslash G/K)$). On the other hand the canonical injection of Ω (furnished with the topology of the spectrum) into $\hat{G}$ is injective and bicontinuous on its image which is open (2.5). Since Kostant has shown that the function $\mathfrak{1}$ is isolated in Ω, we see that ε_G must be isolated in $\hat{G}$. q.e.d.

We are now going to derive a positive result in the cases which have been excluded in the above theorem.

(2.13) Theorem [32] :

If G is simply connected with Lie algebra $so(n;1)$ or $su(n;1)$ then there exist at least one and at most two irreducible representations with a non-trivial cohomology. If there exists one such representation U then it is unitarily equivalent to $\overline{U}$ and $\dim H^1(G,U) \leq 2$. If U is the complexification of an orthogonal representation then $\dim H^1(G,U) = 1$. If there exist two such representations, U_1 and U_2, then $U_1 = \overline{U_2}$ and $\dim H^1(G,U_1) = \dim H^1(G, U_2) = 1$.

Proof:

Because of (2.11) we may restrict ourselves to the case $G = SO_e(n;1)$ or $SU(n;1)$. Let K again denote a maximal compact subgroup of G. Faraut and Harzallah [34] have shown that there exists exactly one nonzero function Q satisfying condition c) of (2.8) (up to multiplication by a real positive scalar) and moreover Q satisfies:

$$\int_K Q(g_1 k g_2)\, dk = Q(g_1) + Q(g_2)$$

Hence if (H_Q, U_Q, δ_Q) is the triple associated with Q we have by (2.7) that U_Q does not contain the trivial representation. Moreover δ_Q which is zero on K is neither a coboundary nor a limit of coboundaries. In fact if we had

$$\delta_Q(g) = \lim_n \{U_Q(g)x_n - x_n\} \quad , \quad x_n \in H_Q$$

(uniformly over compact sets)

$$\text{then } 0 = \int_K \delta_Q(k)\, dk = \lim_n \int_K [U_Q(k)x_n - x_n]\, dk .$$

Since H_Q doesn't contain any vectors invariant under K (apart from the zero vector)

$$\int_K U_Q(k)x_n \, dk = 0 \quad \text{and hence} \quad \lim_{n\to\infty} x_n = 0$$

Then $\delta_Q = 0$ and this gives a contradiction.

Utilizing the direct integral decomposition of U_Q into irreducibles (this is possible, since G is of type I) we obtain that there exists at least one irreducible representation U of G with $H^1(G,U) \neq 0$. Now this U can't be of class I because of (2.6).

Now suppose we have a $\delta \in H^1(G,U)$ and without loss of generality suppose $\delta \equiv 0$ on K. Then we know from (2.10) that by, if necessary, multiplying by a scalar we may assume

$$- \|\delta(g)\|^2 = Q(g) \ .$$

Now our technical lemma (2.2) implies that the subrepresentation V of $U \oplus \overline{U}$ in the subspace E of $H \oplus \overline{H}$ generated by $\delta(g) \oplus \overline{\delta(g)}$ is unitarily equivalent to U_Q since the triple $(E,V,\delta \oplus \overline{\delta})$ is equivalent to (H_Q,U_Q,δ_Q).

If now U' is another irreducible representation (unitary, of course) of G with $H^1(G,U') \neq 0$ then there is a subrepresentation of $U' \oplus \overline{U'}$ which is unitarily equivalent to a subrepresentation of $U \oplus \overline{U}$. Then we easily obtain, using irreducibility, that U is equivalent to U' or $\overline{U}$ is equivalent to U'. Then we have to distinguish two cases:

a) U is not equivalent to U'

Then U_Q, which is its own conjugate as complexification of a real representation is unitarily equivalent to $U \oplus \overline{U}$ and (H_Q,U_Q,δ_Q) is equivalent to $(H \oplus \overline{H}, U \oplus \overline{U}, \overline{\delta' \oplus \delta'})$ if δ' is another element in $Z^1(G,U)$ with $\delta' \equiv 0$ on K then after multiplication by a scalar, if necessary, we have: (H_Q,U_Q,δ_Q) is equivalent to $(H \oplus \overline{H}, U \oplus \overline{U}, \delta' \oplus \overline{\delta'})$. So there exists an operator which commutes with $U \oplus \overline{U}$ and sends $\delta \oplus \overline{\delta}$ to $\delta' \oplus \overline{\delta'}$. Since U is not equivalent to $\overline{U}$ and U is irreducible every operator commuting with $U \oplus \overline{U}$ is of the form $\lambda_1 P_H + \lambda_2 P_{\overline{H}}$ where λ_1, λ_2 are scalars and P_H (respectively $P_{\overline{H}}$) is the projection on H (respectively $\overline{H}$).

Hence $\delta = \lambda\delta'$ for some $\lambda \in$.
Thus $\dim H^1(G,U) = 1$. The same reasoning goes through for $\bar{U}$.

b) U is unitarily equivalent to $\bar{U}$

α) If U_Q is unitarily equivalent to $U \oplus \bar{U}$, as in a), then if δ and δ' are two nontrivial cocycles, which are zero on K , there exists an operator commuting with $U \oplus \bar{U}$ which sends $\delta \oplus \bar{\delta}$ to $\delta' \oplus \bar{\delta'}$.

β) If U_Q , U , and $\bar{U}$ are unitarily equivalent then $(E,V,\delta \oplus \bar{\delta})$ and $(E',V,\delta' \oplus \bar{\delta}')$ (notations as above) are equivalent and V, V' and U are unitarily equivalent. Denoting by W (respectively W') the subrepresentation of $U \oplus \bar{U}$ in $E^{\perp}$ (respectively $E'^{\perp}$) we see that W, W', and U are unitarily equivalent. Hence we may construct an operator C satisfying $C(U \oplus \bar{U}) = (U \oplus \bar{U})C$ which sends $\delta \oplus \bar{\delta}$ to $\delta' \oplus \bar{\delta}'$. Now C may be written as linear combination of $P_H, P_{\bar{H}}, I, J$ where I sends $\bar{H}$ to 0 , and H to $\bar{H}$, and intertwines U and $\bar{U}$. J is defined by $IJ = P_{\bar{H}}$, $JI = P_H$.

Hence in cases α) and β) we have:

$$\delta' = P_H C(\delta+\delta')$$
$$C = \lambda_1 P_H + \lambda_2 P_{\bar{H}} + \lambda_3 I + \lambda_4 J$$

Thus $\delta' = \lambda_1\delta + \lambda_4 J\bar{\delta}$ and so $\dim H^1(G,U) \leq 2$.
If we suppose that U ist the complexification of a representation $U_{\mathbb{R}}$, then $H = H_{\mathbb{R}} + iH_{\mathbb{R}}$. Let $\delta \in Z^1(G,U)$, $\delta \neq 0$, then $\delta = \delta_1 + i\delta_2$ where $\delta_1, \delta_2 \in Z^1(G,U_{\mathbb{R}})$. ($\delta_1, \delta_2, \delta$ zero on K !)

Suppose e.g. that $\delta_1 \neq 0$, then we see easily that (H,U,δ_1) is equivalent to (H_Q,U_Q,δ) . Thus we have case b). One can then choose J intertwining U and $\bar{U}$, such that $J\delta_1 = \delta_1$. Now every other element of $Z^1(G,U)$ may be written as $\delta' = \lambda_1\delta_1 + \lambda_4 J\delta_1 = (\lambda_1 + \lambda_4)\delta_1$; $\lambda_1, \lambda_4 \in \mathbb{C}$. Hence $\dim H^1(G,U) = 1$. q.e.d.

In order to use (2.13) to describe the cohomology of $SO(n;1)$ and $SU(n;1)$ we have to investigate the connection between first order cocycles for the Lie group and first order cocycles for the Lie algebra. In the next section we are going to exhibit some results due to Pinczon and Simon ([23]). This will lead to a computation of the cohomology for $SU(n;1)$ and $SO(n;1)$.

3. The Connection between the Cohomology of Lie Algebra and Lie Group with Applications to SU(n;1) and SO(n;1)

In [23] the computation of $H^1(G,\mathcal{H})$ is first of all reduced to the computation of $H^1(G,\mathcal{H}_\omega)$ where $\mathcal{H}_\omega$ is the space of analytic vectors for the representation in question.

Let G be a connected real Lie group and U be a continuous representation of G in a Banach space $\mathcal{H}$. If $\mathcal{H}_\omega$ is the space of analytic vectors for U we denote by $Z^1_\omega(G,\mathcal{H}_\omega)$ those elements in $Z^1(G,\mathcal{H})$ which are analytic functions on G. It follows from the cocycle identity that the range of such a cocycle is contained in $\mathcal{H}_\omega$. As usual let $B^1(G,\mathcal{H}_\omega) = Z^1(G,\mathcal{H}_\omega) \cap B^1(G,\mathcal{H})$.

With these definitions Pinczon and Simon prove in [23]

(3.1) Lemma:

$$H^1(G,\mathcal{H}) = H^1_\omega(G,\mathcal{H}_\omega)$$

The proof is rather technical and uses properties of increase of solutions of the heat equation.

The point is that with this result one can get a map from $H^1_\omega(G,\mathcal{H}_\omega)$ to $H^1_\omega(\mathfrak{g},\mathcal{H}_\omega)$ where $\mathfrak{g}$ is the Lie algebra of G and the $\mathfrak{g}$-action in $\mathcal{H}_\omega$ is given by the derived representation dU. Thus the connection between $H^1(G,\mathcal{H})$ and $H^1_\omega(\mathfrak{g},\mathcal{H}_\omega)$ becomes obvious.

So suppose we are given an element $\delta \in Z^1(G,\mathcal{H}_\omega)$ and $X \in \mathfrak{g}$. Then we define a map

$$D: Z^1_\omega(G,\mathcal{H}_\omega) \to Z^1(\mathfrak{g},\mathcal{H}_\omega)$$

by $D(\delta)(X) := \frac{d}{dt}[\delta(\exp tX)]_{t=0}$.

It turns out that the kernel of this map is just $B^1(G,\mathcal{H}_\omega)$ and thus we have an induced map of the corresponding cohomology groups.

Using the complexification of a real Lie algebra these results then extend to complex Lie algebras (cf. [23]).

In order to show that the map D described above is in fact a bijection one needs to show that if $\eta \in Z^1(\mathfrak{g},\mathcal{H}_\omega)$ there exists a unique cocycle $\delta \in Z^1_\omega(G,\mathcal{H}_\omega)$ such that $D(\delta) = \eta$. The nontrivial part of this is accomplished by showing that if Π is a representation of $\mathfrak{g}$ defined in a dense invariant domain $\tilde{D}$ in a Banach space $\mathcal{H}$ there exists a unique representation U of G such that $dU|_{\tilde{D}} = \Pi$ (cf. Theorem 4 in [23]). Having done this the required cocycle δ is given by

$$\delta(\exp X) = \int_0^1 e^{t dU(X)} \eta(X)\, dt \quad .$$

Note that here we need to assume G to be simply connected since the above equation defines δ only in a neighbourhood of the identity. But using simple connectedness we may extend to the whole group. (See also I. (2.5).)

In order to compute the cohomology for $SU(n;1)$ and $SO(n;1)$, or rather to make a precise statement about its dimension we need a further result from [23].

Let again $\mathcal{H}_K$ denote the space of all K-finite analytic vectors of U. Let $Z^1_K(\mathfrak{g},\mathcal{H}_K)$, $B^1_K(\mathfrak{g},\mathcal{H}_K)$, respectively $H^1_K(\mathfrak{g},\mathcal{H}_K)$ be the space of cocycles vanishing on $\mathfrak{K}$ (the Lie algebra of K), the space of coboundaries vanishing on $\mathfrak{K}$, respectively $Z^1_K(\mathfrak{g},\mathcal{H}_K)/B^1_K(\mathfrak{g},\mathcal{H}_K)$.

Let $Z^1_K(G,\mathcal{H}_\omega)$, $B^1_K(G,\mathcal{H}_\omega)$, respectively $H^1_K(G,\mathcal{H}_\omega)$ be the space of analytic cocycles vanishing on K, the space of coboundaries vanishing on K, respectively $Z^1_K(G,\mathcal{H}_\omega)/B^1_K(G,\mathcal{H}_\omega)$.

Then we have

(3.2) Lemma [23]:

$$H^1(G,\mathcal{H}_\omega) = H^1_K(G,\mathcal{H}_\omega) \leq H^1_K(\mathfrak{g},\mathcal{H}_K) = H^1(\mathfrak{g},\mathcal{H}_K) \leq H^1(\mathfrak{g},\mathcal{H}_\omega)$$

If G is simply connected all inclusions are equalities.

These results then reduce the global cocycle problem to a study of the problem in the tangent space at the identity. This leads to some powerful theoretical results. However, for the purpose of explicit computation it still seems necessary to use the global approach described for $SL(2;\mathbb{R})$ and $SL(2;\mathbb{C})$.

These results are now sufficient to make a precise statement about the dimension of SU(n;1) for $n \geq 2$ and SO(n;1) for $n \geq 3$. The relevant result is given in [23]. It should be noted that the case SU(1;1) has been dèalt with in extenso in section IV. So we obtain

(3.3) Theorem [32]:

a) If G is a connected Lie group with Lie algebra su(n;1) $(n \geq 2)$ there exist exactly two irreducible unitary representations which are conjugate to each other and have a nontrivial cohomology.

b) If G is a connected Lie group with Lie algebra so(n;1) $n \geq 3$ there exists exactly one irreducible unitary representation (the complexification of a real representation) which has a nontrivial cohomology.

The dimensions of these cohomology groups are all equal to 1 .

Proof:

For the details of the proof we refer the reader to [32]. Let it just be said that (2.12) means that we just need to prove existence of

a) two irreducible unitary representations of G with $H^1 \neq 0$ (conjugate to one another)

b) a unitary irreducible representation of G (complexification of a real representation) with nontrivial cohomology.

This is done in [32] by using induced representations just as in section IV. However, the results are technically very complicated and one doesn't get an explicit realization of the cocycles involved. For this existence proof the results of (3.2) above are of crucial importance since one has to argue via the Lie algebra.

Remark:

In the case of SU(1;1) we can again give explicit results thus tying up the abstract theory of Pinczon/Simon/Delorme with our very concrete calculations:

We simply apply the map D described after (3.1) to our two nontrivial cocycles δ_1 and δ_2 described in IV. (5.4). The cocycles in the Lie algebra associated with the derived representations are then given by

$$D(\delta_1)(X)(x) = (b - ic)x$$

and
$$D(\delta_2)(X)(x) = (b + ic)x$$

where
$$X := \begin{pmatrix} ia & b + ic \\ b - ic & - ia \end{pmatrix} \in su(1;1)$$

This is seen by straightforward computation (Note that one has to be a little careful about taking limits with respect to the correct Hilbert space norm).

VI. "GENUINE" INFINITELY DIVISIBLE REPRESENTATIONS

1. General Definitions

In II.2 we characterized infinitely divisible projective representations in terms of first order cocycles and later on we proceeded to construct CTPs from these. It is clear that if the "multipliers" involved are identically one or trivial (in the sense that they are second order coboundaries) then we arrive at genuine representations. It is, however, not quite clear how this comes about. In this chapter we shall describe infinitely divisible representations (II, Def. 2.1, with all multipliers identically 1) in terms of their "expectation values". These will be related to the "expectation values" of certain projective representations.

(1.1) Definition:

A continuous function $f: G \longrightarrow \mathbb{C}$ is called σ *-positive*, if

$$\sum_{i=1}^{n} \sum_{j=1}^{n} \alpha_i \bar{\alpha}_j \sigma(g_j^{-1}, g_i) f(g_j^{-1} g_i) \geq 0$$

$$\forall\, n \in \mathbb{N}\,,\ \forall\, (\alpha_1, \ldots, \alpha_n) \in \mathbb{C}^n,\ \forall\, (g_1, \ldots, g_n) \in G^n\,.$$

Here σ satisfies the conditions given in II. (1.3)(ii), (iii).

If $\sigma \equiv 1$, then f is positive definite.

Thus a σ-positive function is the obvious generalization of a positive function and indeed we have the example which one would expect:

Example:

Let (U_g, σ, Ω) be a cyclic projective representation. Then the expectation value $f(g) = \langle U_g \Omega, \Omega \rangle$ is σ-positive.

(1.2) Definition:

A pair (f,σ) (with f σ-positive) is called *infinitely divisible*, if $\forall\, n \in \mathbb{N}\ \exists (f_n,\sigma_n)$ with f_n σ_n-positive and $f_n^n = f$ $\sigma_n^n = \sigma$.

Example:

It should be clear that an infinitely divisible cyclic (projective) representation defines an infinitely divisible (σ-) positive function. It should also be clear that the opposite is true, i.e. that every infinitely divisible (σ-)positive function defines an infinitely divisible cyclic (projective) unitary representation up to unitary equivalence.

In II.(2.3) we described all infinitely divisible projective cyclic representations in terms of first order cocycles and thus, of course, all infinitely divisible σ-positive functions. We are now going to exhibit the connection between certain infinitely divisible σ-positive functions and infinitely divisible positive functions for the groups whose cohomology we have studied above. From there it will be obvious how one arrives at CTPs of genuine representations and indeed we shall give some examples later on.

2. Infinitely Divisible Positive Functions for $SO(n) Ⓢ \mathbb{R}^n$, $n \geq 3$

Let $G := SO(n) Ⓢ \mathbb{R}^n$ and let $a: G \to \mathbb{R}$ be continuous with $a(g^{-1}) = -a(g)$ $\forall\, g \in G$.
Then we have the following

(2.1) Lemma:

There exists a bijection Φ between pairs (f,a) with f infinitely divisible positive and certain infinitely divisible σ-positive (Ψ,σ) given by

$\Phi : (f,a) \longmapsto (\Psi,\sigma)$ with

$$\Psi(g) := \exp[i\,a(g)]f(g)$$

$$\sigma(g_1,g_2) := \exp[i(a(g_1)+a(g_2)-a(g_1g_2))].$$

Proof:

(i) Φ is injective:

Let

$$\exp[i\,a_1(g)]f_1(g) = \exp[i\,a_2(g)]f_2(g) \quad \text{and} \quad \sigma_1 \equiv \sigma_2$$

then we obtain

$$f_2(g) = \exp[i(a_1(g)-a_2(g))]f_1(g) .$$

If we set $a_1(g)-a_2(g)=:b(g)$, then $b:G \to \mathbb{R}$ must be a continuous additive homomorphism since $\sigma_1 \equiv \sigma_2$ and continuity of a_1,a_2 implies

$$a_1(g_1)+a_1(g_2)-a_1(g_1g_2) = a_2(g_1)+a_2(g_2)-a_2(g_1g_2) \Rightarrow$$

$$[a_1(g_1)-a_2(g_1)]+[a_1(g_2)-a_2(g_2)] = a_1(g_1g_2)-a_2(g_1g_2) \Rightarrow$$

$$b(g_1g_2) = b(g_1)+b(g_2) \qquad \forall\, g_1,g_2 \in G .$$

But $SO(n)$ is semi-simple $\forall\, n \geq 3$, since the Cartan-Killing form is negative definite. A semi-simple group, however, is its own commutator subgroup. Thus we obtain $b \equiv 0$ on $SO(n)$. A simple computation then shows that $b \equiv 0$ on G . Thus Φ is injective.

(ii) Φ is surjective:

Let (Ψ,σ) be given with

$$\sigma(g_1,g_2) := \exp[i(a(g_1)+a(g_2)-a(g_1g_2))] .$$

We define

$$f(g) := \exp[-ia(g)]\Psi(g) , \quad \text{then}$$

$$\Phi(f,a) = (\Psi,\sigma) .$$

One verifies easily that f is infinitely divisible positive definite $\Leftrightarrow$ $\Phi(f,a)$ is infinitely divisible σ-positive. q.e.d.

Remark:

(2.1) is a generalization of (4.1) in [9].

We are now able to describe the infinitely divisible positive definite functions for the Euclidean Motion Groups. (Strictly speaking we obtain only those which are derived from first order cocycles whose associated representations are irreducible; the general case may be dealt with by considering the decomposition theory given in [20] .) We obtain

(2.2) Theorem:

The infinitely divisible positive definite functions on $SO(n) \circledS \mathbb{R}^n$ are of the form

a) $f(g) = \exp \langle U_g v-v, v\rangle$

b) $f(g) = \exp [-c^2 \|\underline{x}\|^2]$ where

$c \in \mathbb{R}, g=(h,\underline{x}) \in SO(n) \circledS \mathbb{R}^n, g \longmapsto U_g$ is a representation of $SO(n) \circledS \mathbb{R}^n$ and v is some fixed vector in the Hilbert space in which U acts.

Proof:

a) Trivial cocycles (i.e. coboundaries) give the following σ-positive functions according to II. (2.3).
Let $\delta(g)=U_g v-v$, then we have

$$\psi(g) = \exp[\mathrm{Re}\langle U_g v-v,v\rangle+ia(g)]$$

$$\sigma(g_1,g_2) = \exp[i(\alpha(g_1)+\alpha(g_2)-\alpha(g_1g_2))]$$

where $\alpha(g)=a(g)-\mathrm{Im}\langle U_g v,v\rangle$ and $a: SO(n) \circledS \mathbb{R}^n \longrightarrow \mathbb{R}$ is some continuous function satisfying

$$a(g^{-1}) = -a(g) \qquad \forall\, g \in SO(n) \circledS \mathbb{R}^n .$$

An application of lemma (2.1) gives the positive definite functions appearing under a) above.

b) The Maurer-Cartan cocycles give the following σ-positive functions:

$$\Psi(g) = \exp[-c^2 \|\underline{x}\|^2 + ia(g)]$$

$$\sigma(g_1,g_2) = \exp[i(a(g_1)+a(g_2)-a(g_1g_2))] \ .$$

Again using lemma (2.1) we obtain the positive definite functions appearing under b) above. q.e.d.

3. Infinitely Divisible Positive Functions on the Leibniz-Extensions of Certain Compact Lie Groups

For simplicity we consider here compact, connected, semi-simple Lie groups which are real matrix groups.

We do not quite succeed in giving a complete classification of infinitely divisible positive functions on the Leibniz-Extensions $K \circledS k$, K group, k Lie-Algebra, of such groups. The results, however, are not far from complete as we shall see below.

First of all we need an analogue to lemma (2.1). To this end we give the following

(3.1) Definition:

Two pairs (f_1,a_1), (f_2,a_2) as in lemma (2.1) are called equivalent (denoted by $(f_1,a_1) \sim (f_2,a_2)$) if we have:

There exists a continuous, additive homomorphism $b: k \to \mathbb{R}$ with the following properties

(i) $f_2(h,X) = e^{ib(X)} f_1(h,X) \quad \forall (h,X) \in K \circledS k$

(ii) $a_1(h,X) = a_2(h,X) + b(X) \quad \forall (h,X) \in K \circledS k$

(iii) $b(X) = b(\mathrm{Ad}\, h(X)) \quad \forall (h,X) \in K \circledS k$.

One verifies easily that "$\sim$" is an equivalence relation. Let the equivalence class of (f,a) with respect to "$\sim$" be denoted by $\widetilde{(f,a)}$.
Then the analogue to lemma (2.1) is

(3.2) Lemma:

Let $G=K \circledS k$, be the Leibnitz-Extension of a compact, connected, semi-simple, real Lie group. Let further $a:G \to \mathbb{R}$ be continuous with $a(g^{-1}) = -a(g)$, $\forall\, g \in G$. Then there exists a bijection Φ between equivalence classes $\widetilde{(f,a)}$ with f infinitely divisible positive definite and certain infinitely divisible σ-positive (Ψ,σ) given by

$$\Phi: \widetilde{(f,a)} \mapsto (\Psi,\sigma) \qquad \text{with}$$

$$\Psi(g) := \exp[ia(g)]f(g)$$

$$\sigma(g_1,g_2) := \exp[i(a(g_1)+a(g_2)-a(g_1g_2))] \quad .$$

Proof:

First of all one easily verifies that Φ is well-defined.

(i) <u>Φ is injective:</u>

Suppose that

$$\exp[i\, a_1(g)]f_1(g) = \exp[i\, a_2(g)]f_2(g) \quad \text{and} \quad \sigma_1 \equiv \sigma_2 \ .$$

Again set $b:=a_1-a_2$ then by an argument which is analogous to the one used in (2.1) we see that b must be a continuous, additive homomorphism. Since K is by assumption semi-simple again b must be identically zero on K . Thus we may consider b as homomorphism on k which satisfies the condition

$$b(X) = b(\text{Ad}\, h(X)) \qquad \forall (h,X) \in K \circledS k \ .$$

(This is obtained by an easy computation.)

Thus it follows that $(f_1,a_1) \sim (f_2,a_2)$ and that Φ is injective.

(ii) <u>Φ is surjective:</u>

Let (Ψ,σ) be given with

$$\sigma(g_1,g_2) := \exp[i(a(g_1)+a(g_2)-a(g_1g_2))] \ .$$

We then define

$$f(g) := \exp[-ia(g)]\Psi(g) \ .$$

It immediately follows that

$$\Phi(\widetilde{f,a}) = (\Psi,\sigma) .$$

Again one verifies that $\Phi\ (\widetilde{f,a})$ is infinitely divisible σ-positive iff $e^{ib}f$ is infinitely divisible positive for each continuous, additive homomorphism $b:k \to \mathbb{R}$ with

$$b(X) = b(\mathrm{Ad}\ h(X)) \qquad \forall (h,X) \in K \circledS k . \qquad \text{q.e.d.}$$

This lemma leads to

(3.3) Theorem:

The infinitely divisible positive definite functions on $K \circledS k$ are of the form

a) $f(h,X)=\exp[\langle U_{(h,X)}v-v,v\rangle+ib(X)]$

b) $f(h,X)=\exp[ib(X)-c^2\|X\|^2]$, $c \in \mathbb{R}$.

Here b is as in (3.2), $\|\cdot\|^2$ is given by $(-1) \times$ Cartan-Killing form and the interpretation of the other symbols is obvious.

Proof:
The proof is analoguous to the proof of (2.2) and thus we are not going to give the details. Let us remark, however, that the positive definite functions appearing under a) again are obtained by using coboundaries whilst those under b) arise from the Maurer-Cartan cocycles. Again we have only considered cocycles which are associated with irreducible representations. q.e.d.

Remarks:

(i) The classification in (3.3) is not entirely complete since in each special case one has to investigate the existence of a non-trivial homomorphism $b:k \to \mathbb{R}$ with the required properties. We recall that for $SO(3)$ for example there is no such b as we have seen above.

(ii) The connectedness condition could be omitted. This would lead to similar results on certain subgroups. Because of some technical complications we have not dealt with this case.

4. Infinitely Divisible Positive Functions on the First Leibniz-Extension of SL(2; ℝ)

This is the final group which we wish to investigate. A simple analysis of the group action on sl(2; ℝ) and of the associated orbits shows that an exact analogue to (2.2) exists in this case. Thus we obtain

(4.1) Theorem:

The infinitely divisible positive functions on SL(2; ℝ) Ⓢ sl(2; ℝ) which arise from trivial cocycles are given by

$$f(g,X) = \exp\langle U_{(g,X)}v - v, v\rangle \ .$$

For non-trivial cocycles which are associated with irreducible representations there are no infinitely divisible positive definite functions.

Proof:
The first part of the statement should be obvious by now.

The second part arises since the two non-trivial cocycles δ_1, δ_2 as computed in IV give rise to multipliers σ_1, σ_2 which cannot be written as

$$\sigma_j(g_1,g_2) = \exp[i(a_j(g_1)+a_j(g_2)-a_j(g_1 g_2))] \ , \quad j=1,2$$

for some continuous function a_j: SL(2;ℝ) Ⓢ sl(2;ℝ) → ℝ . This is in fact not so easy to see and we refer the reader to [5] for the technical details.

q.e.d.

Remark:

We do, however, obtain an infinitely divisible positive definite function, which comes from a non-trivial cocycle, if we set

$$f(g,X) := \exp[-c^2 \|\delta(g,x)\|^2] , \quad c \in \mathbb{R}$$

where $\delta := \delta_1 + \delta_2$.
It should be noted that the representation associated with δ is no longer irreducible!

5. The Explicit Formula for the Representations

In this section we wish to describe explicitly the representations described above purely in terms of their expectation values. These are not too hard to obtain and we proceed to write down the formula without further ado.

(5.1) $SO(n)$ Ⓢ $\mathbb{R}^n$:

We recall that the infinitely divisible positive definite functions in this case had the form

a) $f(g) = \exp\langle U_g v - v, v\rangle$

b) $f(g) = \exp[-c^2 \|\underline{x}\|^2]$.

The notation is the same as in (2.2) above.

We consider case a) first:
We define a cyclic representation $(V_g, \frac{\operatorname{Exp} v}{\|\operatorname{Exp} v\|})$ in Fock space by means of

$$V_g \operatorname{Exp} x := \operatorname{Exp} U_g x .$$

It is clear that this defines a representation (unitary of course) if we extend by linearity. Furthermore its expectation value is given by

$$\frac{1}{\|\operatorname{Exp} v\|^2}\langle V_g \operatorname{Exp} v, \operatorname{Exp} v\rangle = \exp\langle U_g v - v, v\rangle .$$

Case b):
In this case we define a cyclic representation $(V_{(A,\underline{x})}, \mathrm{Exp}\ 0)$ in Fock space by setting

$$V_{(A,\underline{x})} : \mathrm{Exp}\ \underline{x}' \longmapsto$$

$$\exp[-c^2 \|\underline{x}\|^2 - \sqrt{2}\, c < \underline{x}, A\underline{x}'>]\ \mathrm{Exp}(A\underline{x}' + \sqrt{2}\, c\, \underline{x}) .$$

Then it is easily seen that we again obtain a unitary representation whose expectation value is indeed $\exp[-c^2 \|\underline{x}\|^2]$ as required.

Remark:

It should be noted that if we take $c = \frac{1}{\sqrt{2}}$ in the above formula then we obtain as a special case one of the representations constructed in [11] p. 24 .

(5.2) The Leibnitz-Extension of Certain Compact Groups

We consider the same groups as in section 3 above; the notation will also remain the same. Again there are two cases which are almost analogous to (5.1).

a) The expectation value is given by

$$f(h,X) = \exp[<U_{(h,x)} v - v, v> + ib(X)] .$$

b) The expectation value is given by

$$f(h,X) = \exp[ib(X) - c^2 \|X\|^2] .$$

Case a):
Define a cyclic representation in Fock space $(V_g, \mathrm{Exp}\ \frac{v}{\|v\|})$, where $g=(h,X)$, by setting

$$V_g\ \mathrm{Exp}\ x := e^{ib(X)}\, \mathrm{Exp}\ U_{(h,X)} x .$$

It is a simple computation to show that this defines a representation with the required expectation value.

Case b):
Again the analogues to (5.1) are quite obvious. We define $(V_{(h,X)},$ Exp O) by setting

$$V_{(h,X)} : \mathrm{Exp}\; X' \longmapsto$$
$$\exp[ib(X) - c^2 \|X\|^2 - \sqrt{2}\, c \langle X, \mathrm{Ad}\; h(X')\rangle]\,\mathrm{Exp}(\mathrm{Ad}\; h(X') + \sqrt{2}\, c\, X) \;.$$

Again it is easy to see that this will give the required result.

(5.3) SL(2;$\mathbb{R}$):

The infinitely divisible representations arising from trivial cocycles via infinitely divisible positive definite functions are again constructed as in (5.1). We are not going to elaborate further on this point but are instead going to describe the infinitely divisible projective representations arising from the two cocycles δ_1, δ_2 described in IV as well as the infinitely divisible genuine representations arising from $\delta_1 + \delta_2$.

a) Projective Representation from δ_1 :

Let
$$g = \begin{bmatrix} \alpha & \beta \\ \bar{\beta} & \bar{\alpha} \end{bmatrix} \qquad g' = \begin{bmatrix} \alpha' & \beta' \\ \bar{\beta}' & \bar{\alpha}' \end{bmatrix} \;.$$

Then $(U_g, \exp i\, \mathrm{Im} \langle \delta(g'), \delta(g^{-1})\rangle, \mathrm{Exp}\; O)$ is defined by

$$U_g \mathrm{Exp}\; \delta(g') := \left(\frac{\alpha\alpha' + \beta\beta'}{\alpha\alpha'}\right)^{-\lambda} |\alpha|^{-\lambda} \mathrm{Exp}\; \delta(gg')$$

where the representation is defined in the space spanned by $\{\mathrm{Exp}\; \delta(g) : g \in SU(1,1)\}$ and $\lambda = 8\pi^2 |C|^2$, $\delta = C \cdot \delta_1$.

b) Projective Representations from δ_2 :

Let g,g', as in a) above then $(U_g, \exp i\, \mathrm{Im} \langle \delta(g'), \delta(g^{-1})\rangle,$ Exp O) is defined by

$$U_g \mathrm{Exp}\; \delta(g') := \left(\frac{\alpha\bar{\alpha}' + \bar{\beta}\beta'}{\alpha\alpha}\right)^{-\lambda} |\alpha|^{-\lambda} \mathrm{Exp}\; \delta(gg')$$

where $\delta = C \cdot \delta_2$ and the representation space is again spanned by

$\{\text{Exp } \delta(g) : g \in SU(1,1)\}$.

c) <u>Genuine Representation from $\delta_1 + \delta_2$:</u>

The representation $(U_g, \text{Exp } 0)$ is defined by

$$U_g \text{ Exp } \delta(g') := \left(\frac{\alpha\alpha' + \bar{\beta}\bar{\beta}'}{\alpha\alpha}\right)^{-2\lambda} |x|^{-2\lambda} \text{ Exp } \delta(gg')$$

where $\delta = C(\delta_1 + \delta_2)$.

Remark:

The genuine representation described under c) above is just the tensor product of the two projective representations under a) and b). Their multipliers are complex conjugate to each other and thus the product is identically one.

6. Some Remarks on Irreducibility

An interesting question to answer for the representations described above is whether they are irreducible or not. It is obvious from our discussion of coboundaries in the first chapter that those in the case of infinitely divisible representations also will provide only reducible representations. In general it is not always easy to see whether one of the representations constructed above is irreducible or not (there just doesn't seem to exist an algorithm for dealing with this sort of thing). Thus by way of illustration we are here going to consider the Maurer-Cartan cocycle for $SO(3) \circledS \mathbb{R}^3$ and the nontrivial cocycle $\delta = C(\delta_1 + \delta_2)$ for $SU(1;1)$.

(6.1) $SO(3) \circledS \mathbb{R}^3$:

As pointed out in chapter III. this is a regular semi-direct product and thus we know <u>all</u> irreducible representations. Indeed they are described by III. (1.3).

A close examination shows that all representations are induced from rotations around the z-axis and we have the following precise form:

(i) $(A,\underline{b}) \mapsto V^1_{(A,\underline{b})}$ $\qquad (A,\underline{b}) \in SO(3) \circledS \mathbb{R}^3$

where $(V^{(1)}_{(A,\underline{b})}f)(\underline{x}) = e^{i<\underline{x},\underline{b}>}f(A^{-1}x)$

$\underline{x} \in S_R := \{\underline{y} \in \mathbb{R}^3 : ||\underline{y}||^2 = R^2\}$

$f \in \mathcal{L}^2(S_R, R^2 \sin\theta d\theta d\phi)$

(ii) $(A,\underline{b}) \mapsto V^{(2)}_{(A,\underline{b})}$

where $(V^{(2)}_{(A,\underline{b})}f)(\underline{x}) = e^{i<\underline{x},\underline{b}>}e^{in\theta(A,\underline{x})}f(A^{-1}x)$,

$f,\underline{x}$ are as in (i)

Here θ is some real-valued function $SO(3) \circledS \mathbb{R}^3 \to \mathbb{R}$, $n \in \mathbb{Z}$ and the $e^{in\theta(A,\underline{x})}$- term corresponds to the $C(h,h^{-1}\chi')$-term in III. (1.3.).

It should be noted that the surface measure is invariant under the SO(3)-action and thus neither in (i) nor (ii) above does a Radon-Nykodym derivative appear.

We now consider the following representation in Fock-space (it corresponds to the choice of $C = \frac{1}{\sqrt{2}}$ in the Maurer-Cartan cocycle):

$$U_{(A,\underline{b})} : \text{Exp } \underline{a} \mapsto \exp[\frac{-||b||^2}{2} - <A\underline{a},\underline{b}>]\text{Exp}(A\underline{a} + \underline{b})$$
$$\underline{a} \in \mathbb{R}^3 .$$

If this were irreducible then it would have to be unitarily equivalent to one of the representations listed under (i) respectively (ii) above.

First of all we make two fairly trivial observations (which nevertheless are going to be crucial in the following argument):

(1) $U_{(A,\underline{0})}\text{Exp } \underline{0} = \text{Exp } \underline{0} \qquad \forall A \in SO(3)$

(2) $U_{(I,\underline{b})}\text{Exp } \underline{0} = \exp[\frac{-||b||^2}{2}]\text{Exp } \underline{b} \qquad \forall \underline{b} \in \mathbb{R}^3$.

If our representation U were equivalent to one of the representations listed under (i) above then we would have a unitary operator T with the property

$$T\, U_{(A,\underline{b})} = V^{(1)}_{(A,\underline{b})}\, T \qquad \forall (A,\underline{b}) \in SO(3) \circledS \mathbb{R}^3 .$$

But then consider $T \operatorname{Exp} \underline{0} := f$ say, some fixed function in $\mathcal{L}^2(S_R, R^2 \sin\theta d\theta d\phi)$ for some radius R. Further we would have
$T U_{(A,\underline{0})} \operatorname{Exp} \underline{0} = f \quad \forall A \in SO(3)$

(see observation (1) above), and

$$[V^{(1)}_{(A,\underline{0})} T \operatorname{Exp} \underline{0}](\underline{x}) = f(A^{-1}\underline{x}) .$$

Thus $f(\underline{x}) = f(A^{-1}\underline{x}) \quad \forall A \in SO(3)$ a.e.$\underline{x}$.

The SO(3)-action in all spheres is, however, transitive. Thus we obtain immediately that

$$f(\underline{x}) = \text{constant a.e. } \underline{x}$$

Now

$$T U_{(I,\underline{b})} \operatorname{Exp} \underline{0} = T \exp\left[\frac{-||b||^2}{2}\right] \operatorname{Exp} \underline{b}$$

(see observation (2) above)

and

$$[V^{(1)}_{(I,\underline{b})} T \operatorname{Exp} \underline{0}](\underline{x}) = \text{constant} \times e^{i\langle \underline{x},\underline{b}\rangle}$$

Thus

$$[T \operatorname{Exp} \underline{b}](\underline{x}) = \text{constant} \times e^{i\langle \underline{x},b\rangle + \frac{||b||^2}{2}}$$

So T would be defined (up to a constant) on a set of vectors which span the Fock space.

The fact that T must conserve lengths leads to the requirement that $R = \frac{1}{2\sqrt{\pi}}$ (without loss of generality we have assumed the constant to be one in order to avoid having to write it down in all our calculations). But then one obtains easily that

$$\langle T \operatorname{Exp} \underline{a}, T \operatorname{Exp} \underline{b}\rangle = \exp\left[\tfrac{1}{2}(||\underline{a}||^2 + ||\underline{b}||^2)\right] \frac{\sin||a-b||R}{||a-b||R}$$

with $R = \frac{1}{2\sqrt{\pi}}$.

Thus T can't be unitary and so $U_{(A,\underline{b})}$ can't be unitarily equivalent to one of the representations listed under (i) above. It remains to deal with the representations under (ii). As before suppose that a T exists (unitary, of course) with

$$T\, U_{(A,\underline{b})} = V^{(2)}_{(A,\underline{b})}\, T \qquad \forall (A,\underline{b}) \in SO(3) \circledS \mathbb{R}^3$$

Again set $T \quad \operatorname{Exp} \underline{O} := f$. Then this leads to

$$f(\underline{x}) = e^{in\theta(A,\underline{x})} f(A^{-1}\underline{x}) \qquad \forall\, A \in SO(3), \qquad \text{a.e. } \underline{x} .$$

As above we see easily that f must have constant modulus. Hence

$$e^{in\theta(A,\underline{x})} = f'(\underline{x}) f'(A^{-1}\underline{x})^{-1}$$

for some function f' of modulus one. But then we would be dealing with a representation which would be unitarily equivalent to one of those listed under (i) above, cf.[15] p. 147. Thus our representation can't be irreducible.

(6.2) SU(1;1) :

First of all we note that the infinitely divisible positive definite function constructed in (4.1) from $\delta = C(\delta_1 + \delta_2)$ may be viewed as an infinitely divisible positive function on $SU(1;1)$. If in fact we give an explicit realization we obtain:

For any $\lambda \geq 0$, $g = \begin{pmatrix} \alpha & \beta \\ \bar{\beta} & \bar{\alpha} \end{pmatrix} \in SU(1;1)$

$$\Psi^{\lambda}(g) := |\alpha|^{-2\lambda} = \left[\frac{4}{\operatorname{tr} gg^* + 2}\right]^{\lambda}$$

is an infinitely divisible positive function on $SU(1;1)$. It results from

$$\delta = C(\delta_1 + \delta_2) \qquad \text{with} \qquad \lambda = 8\pi^2 |C|^2$$

Remark:

These are precisely the infinitely divisible positive functions described in [28] p. 90. Positive definiteness is there proved by a rather laborious ad hoc method. We are now going to investigate the irreducibility of the representation with "expectation value" Ψ^{λ}. Theorem (1.2) in [28] tells us:

a) if $\lambda > 1/2$, then the representations are given by direct integrals over representations of the principal series and hence reducible.

b) if $\lambda < 1/2$, then the representations split into a direct sum:

One summand belongs to the complementary series whilst the other is a direct integral of representations in the principal series. Hence these representations are reducible also.

The case $\lambda = 1/2$ requires special consideration. For technical convenience we consider the problem on $SL(2;\mathbb{R})$, using the isomorphism

$$SU(1;1) \ni \begin{pmatrix} \alpha & \beta \\ \bar{\beta} & \bar{\alpha} \end{pmatrix} \overset{\eta}{\longmapsto} \begin{pmatrix} a & b \\ c & d \end{pmatrix} \in SL(2;\mathbb{R})$$

where $\alpha = \frac{1}{2}[(a + d) + i(b - c)]$, $\beta = \frac{1}{2}[(b + c) + i(a - d)]$.

Thus our functions Ψ^{λ} give rise to functions f^{λ} on $SL(2;\mathbb{R})$ by setting

$$f^{\lambda}\left(\begin{pmatrix} a & b \\ c & d \end{pmatrix}\right) := \psi^{\lambda}\left(\eta\left(\begin{pmatrix} \alpha & \beta \\ \bar{\beta} & \bar{\alpha} \end{pmatrix}\right)\right) = \left[\frac{4}{(a+d)^2 + (b-c)^2}\right]^{\lambda}.$$

Now according to theorem 9 in chapter IV. of [13] this "expectation value" defines an irreducible representation iff f^{λ} is a spherical function. This is the case iff

$$\int_K f^{\lambda}(x_1 k x_2)\,dk = f^{\lambda}(x_1) f^{\lambda}(x_2) \qquad \forall\, x_1, x_2 \in SL(2;\mathbb{R}) \qquad (*)$$

where $K := \left\{ \begin{pmatrix} \cos\theta & \sin\theta \\ -\sin\theta & \cos\theta \end{pmatrix} : \theta \in R \right\}$ and dk denotes the normalized Haar measure on K .

We consider the case $\lambda = 1/2$, $x_i := \begin{pmatrix} a_i & 0 \\ 0 & a_i^{-1} \end{pmatrix}$,

$a_i \in \mathbb{R}$, $a_i > 0$, $i = 1,2$.

Let $k = \begin{pmatrix} \cos\theta & \sin\theta \\ -\sin\theta & \cos\theta \end{pmatrix}$, then (*) reduces to

$$\int_0^{2\pi} \frac{(1 + \beta)\, d\theta}{[1 + (\beta^2-1)\cos^2\theta]^{1/2}} = 4\pi$$

where
$$\beta = \frac{1 + (a_1 a_2)^2}{a_1^2 + a_2^2}$$

Set
$$F(\beta) := \int_0^{2\pi} \frac{(1 + \beta)\ \ d\theta}{[1 + (\beta^2-1)\cos^2\theta]^{1/2}},$$ then this defines

a function which is differentiable at $\beta = 1$.

We obtain $F'(1) = -2\pi \neq 0$.

Thus clearly (*) can't be satisfied and the associated representation is (unfortunately) reducible even in this case.

APPENDIX

In this appendix we are going to describe briefly the connection between σ-positive functions on the one hand and projective cyclic unitary representations on the other hand. Suppose a projective continuous unitary representation $g \mapsto U_g$ with $U_{g_1} U_{g_2} = \sigma(g_1,g_2) U_{g_1 g_2}$ is given. Let x, with $\|x\|=1$, be a cyclic vector for this representation. Then we have

Lemma 1:

Let $f(g) := \langle U_g x, x\rangle$ be the expectation value of U_g with respect to x. Then f is σ-positive.

Proof:
The continuity of f is obvious, and we immediately obtain

$$\sum_{i=1}^{n} \sum_{j=1}^{n} \alpha_i \bar{\alpha}_j \, \sigma(g_j^{-1}, g_i) f(g_j^{-1} g_i) =$$

$$\sum_{i=1}^{n} \sum_{j=1}^{n} \alpha_i \bar{\alpha}_j \, \langle \sigma(g_j^{-1}, g_i) U_{g_j^{-1} g_i} x, x\rangle =$$

$$\| \sum_{i=1}^{n} \alpha_i U_{g_i} x \|^2 \geq 0 \; .$$

q.e.d.

Remark:

If $\sigma \equiv 1$ then we obtain the well known fact about positive definite functions.

Rather more interesting is the fact that a converse to Lemma 1 exists. We recall that a central extension G_σ of G is defined by

$$G_\sigma : = G x S^1 \text{ as a set, with}$$

$$(g_1,t_1)\cdot(g_2,t_2) : = (g_1 g_2, \sigma(g_1,g_2) t_1 t_2)$$

where G_σ (cf. I.3) is furnished with the product topology (note that we assumed σ to be continuous!). This central extension allows us to consider "genuine" representations of G_σ instead of projective rep-

resentations of G .
Indeed, let $g \mapsto U_g$ be as above; we set

$$V_{(g,t)} := t\, U_g$$

and obtain as is readily verified a "genuine" representation for G_σ with cyclic vector x . Thus we can state

Lemma 2:

Let f be σ-positive on G . Then there exists a projective representation $g \mapsto U_g$ with $U_{g_1} U_{g_2} = \sigma(g_1,g_2)\, U_{g_1 g_2}$ and cyclic vector x satisfying

$$f(g) = \langle U_g x, x\rangle .$$

Moreover U_g and x are determined up to unitary equivalence.

Proof:
We first set

$$f_1(g,t) := t\, f(g) .$$

Then f_1 is positive definite on G_σ , since we have:

$$\sum_{i=1}^{n} \sum_{j=1}^{n} \alpha_i \bar{\alpha}_j\, f_1((g_j,t_j)^{-1} \cdot (g_i,t_i)) =$$

$$\sum_{i=1}^{n} \sum_{j=1}^{n} \alpha_i \bar{\alpha}_j\, f_1(g_j^{-1} g_i, t_j^{-1} t_i\, \sigma(g_j^{-1},g_i)) =$$

$$\sum_{i=1}^{n} \sum_{j=1}^{n} (\alpha_i t_i)\,(\overline{\alpha_j t_j})\, \sigma(g_j^{-1},g_i)\, f(g_j^{-1} g_i) \geq 0 .$$

(The last inequality follows since f is by assumption σ-positive.)

The well-known theorem concerning positive definite functions tells us now that there exists a representation $(g,t) \mapsto V_{(g,t)}$ of G_σ with cyclic vector x and

$$f_1(g,t) = \langle V_{(g,t)} x, x\rangle$$

where V and x are determined up to unitary equivalence. Since $f_1(g,t)=t\cdot f(g)$ we must have $V_{(g,t)}=t\cdot U_g$ for some unitary operator U_g . One verifies immediately that

$$U_{g_1}U_{g_2} = \sigma(g_1,g_2)U_{g_1g_2} \quad \forall g_1,g_2 .$$ q.e.d.

Remark:

The connection between σ-positive functions and projective representations was probably first recognized by Araki. Since it is of crucial importance in these notes we have given the description here again.

REFERENCES

[1] Araki, H.: Factorizable representation of Current Algebra-non commutative extension of Levy Kinchin formula and cohomology of a solvable group with values in a Hilbert space. RIMS, 41 (Revised Edition), Kyoto (1969), (Preprint)

[2] Bargmann, V.: Irreducible unitary representations of the Lorentz group. Annals of Mathematics 48, (1947), p. 568-640

[3] Delorme, P.: 1-cohomologie et produits tensoriel continus de representations. These, L'Université de Paris VI (1975)

[4] Erven, J.: Über Kozyklen erster Ordnung von $SL(2;\mathbb{R})$. Dissertation, TU München (1979)

[5] Erven, J., Falkowski, B.-J.: A note on the continuous second cohomology group for $SL(2;\mathbb{R})$. (Unpublished)

[6] Falkowski, B.-J.: Factorizable and infinitely divisible PUA representations of locally compact groups. J. of Math. Phys., 15, (1974), p. 1060-1066

[7] Falkowski, B.-J.: First order cocycles for $SL(2;\mathbb{C})$. J. of the Ind. Math. Soc., 41, (1977), p. 245-254

[8] Falkowski, B.-J.: A note on the first order cohomology for $SL(n;\mathbb{C})$. J. of the Ind. Math. Soc., 42, (1978), p. 105-107

[9] Falkowski, B.-J.: Infinitely divisible positive definite functions on $SO(3) \circledS \mathbb{R}^3$. p. 111-115 in: Probability measures on groups. Lecture Notes in Mathematics, Vol. 706, Springer-Verlag, Berlin (1979)

[10] Falkowski, B.-J.: Die Kohomologie gewisser Gruppenerweiterungen und faktorisierbare Darstellungen. Unpublished manuscript, Hochschule der Bundeswehr München (1979)

[11] Guichardet, A.: Symmetric Hilbert spaces and related topics. Lecture Notes in Mathematics, Vol. 261, Springer-Verlag, Berlin (1972)

[12] Helgason, S.: Differential geometry and symmetric spaces. Academic Press, New York and London (1962)

[13] Lang, S.: $SL_2(\mathbb{R})$. Addison-Wesley Publishing Company, Reading, Mass. (1975)

[14] Lipsman, R.L.: Group representations. Lecture Notes in Mathematics, Vol. 388, Springer-Verlag, Berlin (1974)

[15] Mackey, G.W.: Induced representations of groups and quantum mechanics. Benjamin, New York (1968)

[16] Mackey, G.W.: The theory of unitary group representations. Chicago Lecture Notes (unpublished manuscript)

[17] Newman, C.M.: Ultralocal quantum field theory in terms of currents. Comm. in Math. Phys., 26, (1972), p. 169-204

[18] Parry, W., Schmidt, K.: A note on cocycles of unitary representations. Proc. Americ. Math. Soc. 55 (1976), p. 185-190

[19] Parthasarathy, K.R.: Probability measures on metric spaces. Academic Press, New York (1967)

[20] Parthasarathy, K.R., Schmidt, K.: Positive definite kernels, continuous tensor products, and central limit theorems of probability theory. Lecture Notes in Mathematics, Vol. 272, Springer-Verlag, Berlin (1972)

[21] Parthasarathy, K.R., Schmidt, K.: Factorizable representations of current groups and the Araki-Woods imbedding theorem. Acta Math. 128, (1972), p. 53-71

[22] Parthasarathy, K.R., Schmidt, K.: A new method for constructing factorisable representations for current groups and current algebras. Comm. in Math. Phys., 50, (1976), p. 167-175

[23] Pinczon, G., Simon, E.: On the 1-cohomology of Lie groups. Letters in Math. Phys. 1, (1975), p. 83-91

[24] Schmidt, K.: Algebras with quasilocal structure and factorizable representations. p. 237-251 in: Mathematics of Contemporary Physics. Academic Press, New York (1972)

[25] Simms, D.J.: Lie groups and quantum mechanics. Lecture Notes in Mathematics, Vol. 52, Springer-Verlag, Berlin (1968)

[26] Streater, R.F.: Current commutation relations, continuous tensor products and infinitely divisible group representations. Rend. Sci. Int. Fisica E. Fermi, XI, (1969), p. 247-263

[27] Takahashi, R.: Sur les fonctions sphériques et la formule de Plancherel dans le group hyperbolique. Japan J. Math., 31, (1961), p. 55-90

[28] Vershik, A.M., Gelfand, J.M., Graev, M.J.: Representations of the group $SL(2;R)$, where R is a ring of functions. Russ. Math. Surv., 28, (1973), p. 87-132

[29] Vershik, A.M., Gelfand, J.M., Graev, M.J.: Irreducible representations of the group G^X and cohomologies. Funct. Anal. and Appl. (Translation), Vol. 8, No. 2, p. 67-69 (1974)

[30] Warner, G.: Harmonic analysis on semi-simple Lie groups I. Springer-Verlag, Berlin (1972)

[31] Stasheff, J.D.: Continuous cohomology of groups and classifying spaces. Bull. of the AMS, Vol. 84, No. 4, (1978), p. 513 - 530.

[32] Delorme, P.: 1-cohomologie des reprêsentations unitaires des Groupes de Lie semi-simples et résolubles. Bull. Soc. math. France, 105, 1977, p. 281 - 336

[33] Dixmier, J.: C-Algebras, North-Holland, 1977

[34] Faraut/Harzallah: Distances hilbertiennes invariantes sur un espace homogêne. Ann. Inst. Fourier

[35] Gangolli: Positive Definite Kernels on Homogeneous Spaces and Certain Stochastic Processes related to Levy's Brownian Motion of Several Parameters. Ann. I.H.P.B. Vol. III, No. 2 (1967)

[36] Warner, G.: Harmonic analysis on semi-simple Lie groups II. Springer-Verlag, Berlin (1972)

[37] Wang, S.P.: The Dual of Semi-simple Lie Groups. Am. J. Math. (1969), p. 921 - 937